Formidable Problems in Electromagnetism

(FULLY SOLVED)

Dr. Sunirmit Verma

ISBN

Hardcase 979-8-89673-017-0
Paperback 979-8-89673-016-3

Contents

What this Book is About

This book is a collection of convoluted problems created by me in the field of electromagnetism. By convoluted, I mean they are modifications of problems found in standard physics textbooks on the topic which require more rigorous mathematical treatment. There are a total of 30 problems, and all of them involve mathematical expression generation, that is, some parameters are given in question in symbolic form instead of numerical values, and what is required to be found out is to be done in the form of a mathematical expression in terms of mentioned parameters. All the problems are fully solved with every step of the solution explained meticulously. The aim of the book is to help readers develop mathematical skills to solve problems on the subject

Acknowledgement

I dedicate this book to my parents, who consecrated their lives to the sole aim of filling my stay in this Earthly realm with exuberance and shielding me from all sorrows out there.

Question 1

At $t=0$, a charged particle having charge q and mass m is launched from the origin with an initial velocity vector: $\vec{v}_{initial} = v_{x0}\hat{i} + v_{y0}\hat{j}$ in a region having a magnetic field $\vec{B} = B_1\hat{i} + B_2\hat{j}$. Find position vector of the particle at time t. Consider only electromagnetic forces.

SOLUTION

Applying Newton's second law of motion on the particle at a general time t

$$\vec{F}_{net} = m(\vec{acc})$$

$$\Rightarrow q\vec{v}\times\vec{B} = m(\vec{acc})$$

$$\Rightarrow q\left(v_x\hat{i} + v_y\hat{j} + v_z\hat{k}\right)\times\left(B_1\hat{i} + B_2\hat{j}\right) = m\left(\frac{dv_x}{dt}\hat{i} + \frac{dv_y}{dt}\hat{j} + \frac{dv_z}{dt}\hat{k}\right) \tag{1}$$

Finding the cross product

$$-\frac{qv_zB_2}{m}\hat{i} + \frac{qv_zB_1}{m}\hat{j} + \frac{q}{m}\left(v_xB_2 - v_yB_1\right)\hat{k} = \frac{dv_x}{dt}\hat{i} + \frac{dv_y}{dt}\hat{j} + \frac{dv_z}{dt}\hat{k} \tag{2}$$

Equating the vector components

$$\frac{dv_x}{dt} = -\frac{qv_z B_2}{m} \tag{3}$$

$$\frac{dv_y}{dt} = \frac{qv_z B_1}{m} \tag{4}$$

$$\frac{dv_z}{dt} = \frac{q}{m}\left(v_x B_2 - v_y B_1\right) \tag{5}$$

Equations 3,4,5 are intercoupled ordinary differential equations in $v_x(t)$, $v_y(t)$, $v_z(t)$

Feeding v_z from equation 4 in 3 and 5

$$\frac{dv_x}{dt} = -\frac{B_2}{B_1}\frac{dv_y}{dt} \tag{6}$$
$$\Rightarrow B_1 dv_x = -B_2 dv_y$$

$$\frac{d}{dt}\left(\frac{m}{qB_1}\frac{dv_y}{dt}\right) = \frac{q}{m}\left(v_x B_2 - v_y B_1\right)$$
$$\Rightarrow \frac{d^2 v_y}{dt^2} = \frac{q^2 B_1}{m^2}\left(v_x B_2 - v_y B_1\right) \tag{7}$$

Integrating both sides in equation 6

$$\int B_1 dv_x = \int -B_2 dv_y$$
$$B_1 v_x + B_2 v_y = C_0 \tag{8}$$

Where C_0 is an arbitrary integration constant. Feeding v_x from equation 8 in 7

$$\frac{d^2 v_y}{dt^2} = \frac{q^2 B_1 B_2}{m^2}\left(\frac{C_0 - B_2 v_y}{B_1}\right) - \frac{q^2 B_1^2}{m^2} v_y$$

$$\Rightarrow \frac{d^2 v_y}{dt^2} + w^2 v_y = \frac{q^2 B_2 C_0}{m^2} \tag{9}$$

Where $w = \dfrac{q\sqrt{B_1^2 + B_2^2}}{m}$. Equation 9 is a 2nd order, linear, inhomogenous, ordinary differential equation with constant coefficients. Its general solution can be given by the method of variation of parameters and is

$$v_y = C_1 \cos wt + C_2 \sin wt + \frac{q^2 B_2 C_0}{m^2 w^2} \tag{10}$$

Where C_1, C_2 are arbitrary integration constants

Feeding equation 10 in 8 and 4

$$v_x = \frac{C_0 - B_2\left(C_1 \cos wt + C_2 \sin wt + \dfrac{q^2 B_2 C_0}{m^2 w^2}\right)}{B_1} \tag{11}$$

$$v_z = \frac{mw}{qB_1}\left[C_2 \cos wt - C_1 \sin wt\right] \tag{12}$$

Equations 10, 11, 12 can be expressed as

$$\frac{dy}{dt} = C_1 \cos wt + C_2 \sin wt + \frac{q^2 B_2 C_0}{m^2 w^2} \tag{13}$$

$$\frac{dx}{dt} = \frac{C_0 - B_2\left(C_1 \cos wt + C_2 \sin wt + \dfrac{q^2 B_2 C_0}{m^2 w^2}\right)}{B_1} \tag{14}$$

$$\frac{dz}{dt} = \frac{mw}{qB_1}\left[C_2 \cos wt - C_1 \sin wt\right] \tag{15}$$

Separating the variables and integrating both sides in each of these equations

$$\int_0^y dy = \int_0^t C_1 \cos wt \cdot dt + \int_0^t C_2 \sin wt \cdot dt + \int_0^t \frac{q^2 B_2 C_0}{m^2 w^2} \cdot dt$$

$$\Rightarrow y = \frac{C_1}{w}\sin wt + \frac{C_2}{w}\left(1 - \cos wt\right) + \frac{q^2 B_2 C_0 t}{m^2 w^2} \tag{16}$$

$$\int_0^x dx = \int_0^t \frac{C_0 - B_2\left(C_1 \cos wt + C_2 \sin wt + \dfrac{q^2 B_2 C_0}{m^2 w^2}\right)}{B_1} \cdot dt$$

$$\Rightarrow x = \frac{C_0 t - \dfrac{B_2 C_1}{w}\sin wt - \dfrac{B_2 C_2\left(1 - \cos wt\right)}{w} - \dfrac{q^2 B_2^2 C_0}{m^2 w^2} t}{B_1} \tag{17}$$

$$\int_0^z dz = \int_0^t \frac{mw\left(C_2 \cos wt - C_1 \sin wt\right)}{qB_1} \cdot dt$$

$$\Rightarrow z = \frac{m\left[C_2 \sin wt - C_1\left(1 - \cos wt\right)\right]}{qB_1} \tag{18}$$

Expressions for x,y,z,v_x,v_y,v_z are now available in terms of time. But they contain unknown constants C_0,C_1,C_2. These need to be found using initial conditions: v_x,v_y,v_z at $t{=}0$ are $v_{x0},v_{y0},0$

$$v_{y0} = C_1 + \frac{q^2 B_2 C_0}{m^2 w^2} \tag{19}$$

$$v_{x0} = \frac{C_0 - B_2\left(C_1 + \frac{q^2 B_2 C_0}{m^2 w^2}\right)}{B_1} \tag{20}$$

$$0 = \frac{mwC_2}{qB_1} \tag{21}$$

Solving,

$$C_0 = B_1 v_{x0} + B_2 v_{y0} \tag{22}$$

$$C_1 = v_{y0} - \frac{q^2 B_2 \left(B_1 v_{x0} + B_2 v_{y0}\right)}{m^2 w^2} \tag{23}$$

$$C_2 = 0 \tag{24}$$

The final position vector will be initial position vector plus displacement vector

$$\vec{r_2} = \vec{r_1} + \vec{s}$$
$$\Rightarrow \vec{r_2} = \left(0\hat{i}+0\hat{j}+0\hat{k}\right)+\left(x\hat{i}+y\hat{j}+z\hat{k}\right) = x\hat{i}+y\hat{j}+z\hat{k} \tag{25}$$

Using equations 16,17,18,22,23,24, equation 25 becomes

$$\vec{r_2} = \left[\frac{\left(B_1 v_{x0} + B_2 v_{y0}\right)\left(1 - \dfrac{q^2 B_2^2}{m^2 w^2}\right)t - \dfrac{B_2}{w}\left\{v_{y0} - \dfrac{q^2 B_2\left(B_1 v_{x0} + B_2 v_{y0}\right)}{m^2 w^2}\right\}\sin wt}{B_1} \right]\hat{i}$$

$$+ \left[\frac{\left\{v_{y0} - \dfrac{q^2 B_2\left(B_1 v_{x0} + B_2 v_{y0}\right)}{m^2 w^2}\right\}}{w}\sin wt + \frac{q^2 B_2\left(B_1 v_{x0} + B_2 v_{y0}\right)t}{m^2 w^2} \right]\hat{j}$$

$$+ \left[-\left\{v_{y0} - \frac{q^2 B_2\left(B_1 v_{x0} + B_2 v_{y0}\right)}{m^2 w^2}\right\}\frac{m(1 - \cos wt)}{q B_1} \right]\hat{k} \tag{26}$$

Expanding back w

$$\vec{r_2} = \left[\frac{B_1\left(B_1 v_{x0} + B_2 v_{y0}\right)t}{B_1^2 + B_2^2} - \frac{mB_2}{qB_1\sqrt{B_1^2 + B_2^2}}\left\{v_{y0} - \frac{B_2\left(B_1 v_{x0} + B_2 v_{y0}\right)}{B_1^2 + B_2^2}\right\}\sin\left(\frac{qt\sqrt{B_1^2 + B_2^2}}{m}\right) \right]\hat{i}$$

$$+ \left[\frac{m}{q\sqrt{B_1^2 + B_2^2}}\left\{v_{y0} - \frac{B_2\left(B_1 v_{x0} + B_2 v_{y0}\right)}{B_1^2 + B_2^2}\right\}\sin\left(\frac{qt\sqrt{B_1^2 + B_2^2}}{m}\right) + \frac{B_2\left(B_1 v_{x0} + B_2 v_{y0}\right)t}{B_1^2 + B_2^2} \right]\hat{j}$$

$$+ \left[-\frac{m}{qB_1}\left\{v_{y0} - \frac{B_2\left(B_1 v_{x0} + B_2 v_{y0}\right)}{B_1^2 + B_2^2}\right\}\left\{1 - \cos\left(\frac{qt\sqrt{B_1^2 + B_2^2}}{m}\right)\right\} \right]\hat{k} \tag{27}$$

Question 2

At $t=0$, a charged particle having charge q and mass m is launched from the origin with an initial velocity vector: $\vec{v}_{initial} = v_0\hat{i}$ in a region having a magnetic field $\vec{B} = B_0\hat{j}$ and an electric field $\vec{E} = E_0 e^{-\lambda t}\hat{k}$. Find position vector of the particle at time t. Consider only electromagnetic forces.

SOLUTION

Applying Newton's second law of motion on the particle at a general time t

$$\vec{F}_{net} = m(\vec{acc})$$
$$\Rightarrow q\vec{v}\mathrm{x}\vec{B} + q\vec{E} = m(\vec{acc}) \tag{1}$$
$$\Rightarrow q\left[\left(v_x\hat{i}+v_y\hat{j}+v_z\hat{k}\right)\mathrm{x}B_0\hat{j}\right] + qE_0 e^{-\lambda t}\hat{k} = m\left(a_x\hat{i}+a_y\hat{j}+a_z\hat{k}\right)$$

Finding the cross product

$$q\left(v_x B_0\hat{k}\text{-}v_z B_0\hat{i}\right) + qE_0 e^{-\lambda t}\hat{k} = m\left(a_x\hat{i}+a_y\hat{j}+a_z\hat{k}\right) \tag{2}$$

Equating the vector components on left and right-hand sides of equation 2

$$a_x = -\frac{q v_z B_0}{m} \tag{3}$$

$$\Rightarrow \frac{dv_x}{dt} = -\frac{q v_z B_0}{m}$$

$$a_z = \frac{q\left(v_x B_0 + E_0 e^{-\lambda t}\right)}{m}$$

$$\Rightarrow \frac{dv_z}{dt} = \frac{q\left(v_x B_0 + E_0 e^{-\lambda t}\right)}{m} \tag{4}$$

Feeding v_x from equation 4 in 3

$$\frac{d}{dt}\left(\frac{\dfrac{m}{q}\dfrac{dv_z}{dt} - E_0 e^{-\lambda t}}{B_0}\right) = -\frac{q v_z B_0}{m} \tag{5}$$

$$\Rightarrow \frac{d^2 v_z}{dt^2} + w^2 v_z = -\frac{q}{m} E_0 \lambda e^{-\lambda t}$$

Where $w = \dfrac{q B_0}{m}$. Equation 5 is a second order linear, inhomogeneous ordinary differential equation with constant coefficients. Its general solution can be given by the method of variation of parameters as

$$v_z = C_1 \cos wt + C_2 \sin wt + \sin wt \int \frac{\left(-\dfrac{q}{m} E_0 \lambda e^{-\lambda t} \cos wt\right)\cdot dt}{w} - \cos wt \int \frac{\left(-\dfrac{q}{m} E_0 \lambda e^{-\lambda t} \sin wt\right)\cdot dt}{w} \tag{6}$$

$$\Rightarrow v_z = C_1 \cos wt + C_2 \sin wt - \frac{q E_0 \lambda \sin wt}{mw}\int e^{-\lambda t} \cos wt \cdot dt + \frac{q E_0 \lambda \cos wt}{mw}\int e^{-\lambda t} \sin wt \cdot dt$$

The two integrals appearing on the right-hand side of equation 6 can be evaluated using integration by parts as illustrated beneath

Let $I_1 = \int e^{-\lambda t} \cos wt \cdot dt$

$$\Rightarrow I_1 = \frac{\cos wt \cdot e^{-\lambda t}}{-\lambda} - \int (-w\sin wt) \cdot \left(\frac{e^{-\lambda t}}{-\lambda} \right) dt$$

$$\Rightarrow I_1 = -\frac{\cos wt \cdot e^{-\lambda t}}{\lambda} - \frac{w}{\lambda} \int \sin wt \cdot e^{-\lambda t} dt$$

$$\Rightarrow I_1 = -\frac{\cos wt \cdot e^{-\lambda t}}{\lambda} - \frac{w}{\lambda} \left[\frac{\sin wt \cdot e^{-\lambda t}}{-\lambda} - \int (w\cos wt) \cdot \left(\frac{e^{-\lambda t}}{-\lambda} \right) dt \right] \qquad (7)$$

$$\Rightarrow I_1 = -\frac{\cos wt \cdot e^{-\lambda t}}{\lambda} - \frac{w}{\lambda} \left[-\frac{\sin wt \cdot e^{-\lambda t}}{\lambda} + \frac{w}{\lambda} I_1 \right]$$

$$\Rightarrow I_1 = -\frac{\cos wt \cdot e^{-\lambda t}}{\lambda} + \frac{w\sin wt \cdot e^{-\lambda t}}{\lambda^2} - \frac{w^2}{\lambda^2} I_1$$

$$\Rightarrow I_1 = \frac{(w\sin wt - \lambda \cos wt)e^{-\lambda t}}{\lambda^2 + w^2}$$

Let $I_2 = \int e^{-\lambda t} \sin wt \cdot dt$

$$\Rightarrow I_2 = \frac{\sin wt \cdot e^{-\lambda t}}{-\lambda} - \int (w\cos wt) \cdot \left(\frac{e^{-\lambda t}}{-\lambda} \right) dt$$

$$\Rightarrow I_2 = -\frac{\sin wt \cdot e^{-\lambda t}}{\lambda} + \frac{w}{\lambda} \int \cos wt \cdot e^{-\lambda t} dt$$

$$\Rightarrow I_2 = -\frac{\sin wt \cdot e^{-\lambda t}}{\lambda} + \frac{w}{\lambda} \left[\frac{\cos wt \cdot e^{-\lambda t}}{-\lambda} - \int (-w\sin wt) \cdot \left(\frac{e^{-\lambda t}}{-\lambda} \right) dt \right] \qquad (8)$$

$$\Rightarrow I_2 = -\frac{\sin wt \cdot e^{-\lambda t}}{\lambda} + \frac{w}{\lambda} \left[-\frac{\cos wt \cdot e^{-\lambda t}}{\lambda} - \frac{w}{\lambda} I_2 \right]$$

$$\Rightarrow I_2 = -\frac{\sin wt \cdot e^{-\lambda t}}{\lambda} - \frac{w\cos wt \cdot e^{-\lambda t}}{\lambda^2} - \frac{w^2}{\lambda^2} I_2$$

$$\Rightarrow I_2 = -\frac{(\lambda \sin wt + w\cos wt)e^{-\lambda t}}{\lambda^2 + w^2}$$

Feeding equations 7 and 8 in 6

$$v_z = C_1 \cos wt + C_2 \sin wt - \frac{qE_0 \lambda e^{-\lambda t}}{m\left(\lambda^2 + w^2\right)} \tag{9}$$

Feeding equation 9 in 4

$$-wC_1 \sin wt + wC_2 \cos wt + \frac{qE_0 \lambda^2 e^{-\lambda t}}{m\left(\lambda^2 + w^2\right)} = \frac{q\left(v_x B_0 + E_0 e^{-\lambda t}\right)}{m}$$

$$\Rightarrow v_x = -\frac{mwC_1}{qB_0}\sin wt + \frac{mwC_2}{qB_0}\cos wt - \frac{E_0 w^2 e^{-\lambda t}}{B_0\left(\lambda^2 + w^2\right)} \tag{10}$$

The constants C_1 and C_2 need to be found using the initial conditions: $v_z\left(t=0\right)=0$, $v_x\left(t=0\right)=v_0$. Using these conditions in equations 9 and 10

$$C_1 = \frac{qE_0 \lambda}{m\left(\lambda^2 + w^2\right)} \tag{11}$$

$$\frac{mwC_2}{qB_0} - \frac{E_0 w^2}{B_0\left(\lambda^2 + w^2\right)} = v_0$$

$$\Rightarrow C_2 = \frac{qB_0}{mw}\left[\frac{E_0 w^2}{B_0\left(\lambda^2 + w^2\right)} + v_0\right] \tag{12}$$

Feeding equations 11 and 12 in 9 and 10

$$v_z = \frac{q}{m\left(\lambda^2 + w^2\right)}\left[E_0 \lambda \cos wt + E_0 w \sin wt + \frac{v_0 B_0\left(\lambda^2 + w^2\right)\sin wt}{w} - E_0 \lambda e^{-\lambda t}\right] \tag{13}$$

$$v_x = \frac{1}{\lambda^2 + w^2}\left[-\frac{E_0 w \lambda}{B_0}\sin wt + \frac{E_0 w^2 \cos wt}{B_0} + v_0\left(\lambda^2 + w^2\right)\cos wt - \frac{E_0 w^2 e^{-\lambda t}}{B_0}\right] \tag{14}$$

Using differential definitions of velocity components, separating the variables and integrating both sides

$$\frac{dz}{dt} = \frac{q\left[E_0\lambda\cos wt + E_0 w\sin wt + \dfrac{v_0 B_0\left(\lambda^2 + w^2\right)\sin wt}{w} - E_0\lambda e^{-\lambda t}\right]}{m\left(\lambda^2 + w^2\right)}$$

$$\Rightarrow \int_0^z dz = \int_0^t \frac{q\left[E_0\lambda\cos wt + E_0 w\sin wt + \dfrac{v_0 B_0\left(\lambda^2 + w^2\right)\sin wt}{w} - E_0\lambda e^{-\lambda t}\right]}{m\left(\lambda^2 + w^2\right)}\, dt \qquad (15)$$

$$\Rightarrow z = \frac{q}{m\left(\lambda^2 + w^2\right)}\left[\frac{E_0\lambda}{w}\sin wt - E_0\cos wt - \frac{v_0 B_0\left(\lambda^2 + w^2\right)\cos wt}{w^2} + E_0 e^{-\lambda t}\right]_0^t$$

$$\Rightarrow z = \frac{q}{m\left(\lambda^2 + w^2\right)}\left[\frac{E_0\lambda}{w}\sin wt + \left(1-\cos wt\right)\left\{ E_0 + \frac{v_0 B_0\left(\lambda^2 + w^2\right)}{w^2}\right\} - E_0\left(1 - e^{-\lambda t}\right)\right]$$

$$\frac{dx}{dt} = \frac{\left[\dfrac{E_0 w^2\cos wt}{B_0} - \dfrac{w E_0\lambda\sin wt}{B_0} + v_0\left(\lambda^2 + w^2\right)\cos wt - \dfrac{E_0 w^2 e^{-\lambda t}}{B_0}\right]}{\lambda^2 + w^2}$$

$$\Rightarrow \int_0^x dx = \int_0^t \frac{\left[\dfrac{E_0 w^2\cos wt}{B_0} - \dfrac{w E_0\lambda\sin wt}{B_0} + v_0\left(\lambda^2 + w^2\right)\cos wt - \dfrac{E_0 w^2 e^{-\lambda t}}{B_0}\right]}{\lambda^2 + w^2}\, dt \qquad (16)$$

$$\Rightarrow x = \frac{\left[\dfrac{E_0 w\sin wt}{B_0} + \dfrac{E_0\lambda\cos wt}{B_0} + \dfrac{v_0\left(\lambda^2 + w^2\right)}{w}\sin wt + \dfrac{E_0 w^2 e^{-\lambda t}}{\lambda B_0}\right]_0^t}{\lambda^2 + w^2}$$

$$\Rightarrow x = \frac{E_0 w\sin wt - E_0\lambda\left(1 - \cos wt\right) + \dfrac{v_0 B_0\left(\lambda^2 + w^2\right)}{w}\sin wt - \dfrac{E_0 w^2\left(1 - e^{-\lambda t}\right)}{\lambda}}{B_0\left(\lambda^2 + w^2\right)}$$

The final position vector will be initial position vector plus displacement vector.

$$\vec{r}_2 = \vec{r}_1 + \vec{s}$$

$$\Rightarrow \vec{r}_2 = \left(0\hat{i}+0\hat{j}+0\hat{k}\right)+\left(x\hat{i}+y\hat{j}+z\hat{k}\right) = x\hat{i}+y\hat{j}+z\hat{k}$$

$$\Rightarrow \vec{r}_2 = \left[\frac{E_0 w\sin wt - E_0\lambda\left(1-\cos wt\right)+\dfrac{v_0 B_0\left(\lambda^2+w^2\right)}{w}\sin wt - \dfrac{E_0 w^2\left(1-e^{-\lambda t}\right)}{\lambda}}{B_0\left(\lambda^2+w^2\right)}\right]\hat{i}+$$

$$\left[\frac{q}{m\left(\lambda^2+w^2\right)}\left[\frac{E_0\lambda}{w}\sin wt + \left(1-\cos wt\right)\left\{E_0 + \frac{v_0 B_0\left(\lambda^2+w^2\right)}{w^2}\right\}-E_0\left(1-e^{-\lambda t}\right)\right]\right]\hat{k} \tag{17}$$

Expanding back w

$$\vec{r}_2 = \left[\frac{\dfrac{E_0 qB_0}{m}\sin\left(\dfrac{qB_0 t}{m}\right)-E_0\lambda\left\{1-\cos\left(\dfrac{qB_0 t}{m}\right)\right\}+\dfrac{mv_0}{q}\left(\lambda^2+\dfrac{q^2 B_0^2}{m^2}\right)\sin\left(\dfrac{qB_0 t}{m}\right)-\dfrac{E_0 q^2 B_0^2\left(1-e^{-\lambda t}\right)}{m^2\lambda}}{B_0\left(\lambda^2+\dfrac{q^2 B_0^2}{m^2}\right)}\right]\hat{i}+$$

$$\left[\frac{\dfrac{E_0\lambda}{B_0}\sin\left(\dfrac{qB_0 t}{m}\right)+\dfrac{q}{m}\left\{1-\cos\left(\dfrac{qB_0 t}{m}\right)\right\}\left\{E_0+\dfrac{m^2 v_0\left(\lambda^2+\dfrac{q^2 B_0^2}{m^2}\right)}{q^2 B_0}\right\}-\dfrac{qE_0}{m}\left(1-e^{-\lambda t}\right)}{\lambda^2+\dfrac{q^2 B_0^2}{m^2}}\right]\hat{k} \tag{18}$$

Question 3

In an imaginary universe, the force vector experienced by a charge q in a magnetic field $\vec{B}$ is given by $\vec{F} = q(\vec{v} \cdot \vec{B})\vec{B}$ where $\vec{v}$ is the instantaneous velocity vector of the particle. At $t=0$, a charged particle having charge q and mass m is launched from the origin with an initial velocity vector: $\vec{v}_{initial} = v_0\hat{i}$ in a region having magnetic field $\vec{B} = B_1\hat{i} + B_2\hat{j}$ and electric field: $\vec{E} = E_0 t\hat{j}$ Find position vector of the particle at time t. Consider only electromagnetic forces.

SOLUTION

Applying Newton's second law of motion on the particle at a general time t

$$\vec{F}_{net} = m(\vec{acc})$$

$$\Rightarrow q(\vec{v} \cdot \vec{B})\vec{B} + q\vec{E} = m(\vec{acc}) \tag{1}$$

$$\Rightarrow q\left[\left(v_x\hat{i} + v_y\hat{j} + v_z\hat{k}\right) \cdot \left(B_1\hat{i} + B_2\hat{j}\right)\right]\left(B_1\hat{i} + B_2\hat{j}\right) + qE_0 t\hat{j} = m\left(a_x\hat{i} + a_y\hat{j} + a_z\hat{k}\right)$$

Finding the dot product

$$q\left(v_x B_1 + v_y B_2\right)\left(B_1\hat{i} + B_2\hat{j}\right) + qE_0 t\hat{j} = m\left(a_x\hat{i} + a_y\hat{j} + a_z\hat{k}\right) \tag{2}$$

Equating the vector components on the left and right-hand sides

$$a_x = \frac{qB_1}{m}\left(v_x B_1 + v_y B_2\right)$$

$$\Rightarrow \frac{dv_x}{dt} = \frac{qB_1}{m}\left(v_x B_1 + v_y B_2\right) \tag{3}$$

$$a_y = \frac{q\left(v_x B_1 B_2 + v_y B_2^2 + E_0 t\right)}{m}$$

$$\Rightarrow \frac{dv_y}{dt} = \frac{q\left(v_x B_1 B_2 + v_y B_2^2 + E_0 t\right)}{m} \tag{4}$$

$$a_z = 0$$

$$\Rightarrow \frac{dv_z}{dt} = 0$$

$$\Rightarrow \int_0^{v_z} dv_z = \int_0^t 0 \cdot dt \tag{5}$$

$$\Rightarrow v_z = 0$$

Solution to equation 5 is pretty straightforward and has been written. Equations 3 and 4 are coupled ordinary differential equations. One of the 2 dependent variables (out of v_x and v_y) needs to be eliminated to proceed further. Feeding v_y from equation 3 in 4

$$\frac{\dfrac{m}{qB_1}\dfrac{d^2 v_x}{dt^2} - \dfrac{dv_x}{dt}}{B_2}B_1 = \frac{\dfrac{mB_2}{B_1}\dfrac{dv_x}{dt} + qE_0 t}{m} \tag{5}$$

$$\Rightarrow \frac{d^2 v_x}{dt^2} - \frac{q\left(B_1^2 + B_2^2\right)}{m}\frac{dv_x}{dt} = \frac{q^2 B_1 B_2 E_0 t}{m^2}$$

Equation 5 is a second order, linear, inhomogenous ordinary differential equation with constant coefficients. Its general solution can be given by the method of variation of parameters and is

$$v_x = C_1 + C_2 e^{\alpha t} + e^{\alpha t}\int \frac{\dfrac{q^2 B_1 B_2 E_0 t}{m^2}}{\alpha e^{\alpha t}}\,dt - \int e^{\alpha t}\,\frac{\dfrac{q^2 B_1 B_2 E_0 t}{m^2}}{\alpha e^{\alpha t}}\,dt \tag{6}$$

Simplifying

$$v_x = C_1 + C_2 e^{\alpha t} + \frac{q^2 B_1 B_2 E_0}{\alpha m^2} e^{\alpha t}\int t e^{-\alpha t}\,dt - \frac{q^2 B_1 B_2 E_0}{\alpha m^2}\int t\cdot dt$$

$$\Rightarrow v_x = C_1 + C_2 e^{\alpha t} + \frac{q^2 B_1 B_2 E_0}{\alpha m^2} e^{\alpha t}\left(-\frac{t e^{-\alpha t}}{\alpha} - \int \frac{e^{-\alpha t}}{-\alpha}\,dt\right) - \frac{q^2 B_1 B_2 E_0 t^2}{2\alpha m^2}$$

$$\Rightarrow v_x = C_1 + C_2 e^{\alpha t} + \frac{q^2 B_1 B_2 E_0}{\alpha m^2} e^{\alpha t}\left(-\frac{t e^{-\alpha t}}{\alpha} - \frac{e^{-\alpha t}}{\alpha^2}\right) - \frac{q^2 B_1 B_2 E_0 t^2}{2\alpha m^2} \tag{7}$$

$$\Rightarrow v_x = C_1 + C_2 e^{\alpha t} - \frac{q^2 B_1 B_2 E_0\left(\alpha t + 1\right)}{\alpha^3 m^2} - \frac{q^2 B_1 B_2 E_0 t^2}{2\alpha m^2}$$

Feeding equation 7 in 3

$$v_y = \frac{m C_2 \alpha e^{\alpha t}}{q B_1 B_2} - \frac{q E_0}{\alpha^2 m} - \frac{q E_0 t}{\alpha m} - C_1 \frac{B_1}{B_2} - C_2 \frac{B_1}{B_2} e^{\alpha t} + \frac{q^2 B_1^2 E_0\left(\alpha t + 1\right)}{\alpha^3 m^2} + \frac{q^2 B_1^2 E_0 t^2}{2\alpha m^2} \tag{8}$$

The constants C_1 and C_2 need to be found using the initial conditions $\left(v_x\right)_{t=0} = v_0$, $\left(v_y\right)_{t=0} = 0$

$$C_1 + C_2 = v_0 + \frac{q^2 B_1 B_2 E_0}{\alpha^3 m^2} \tag{9}$$

$$C_1 \frac{B_1}{B_2} + C_2\left(\frac{B_1}{B_2} - \frac{m\alpha}{q B_1 B_2}\right) = \frac{q^2 B_1^2 E_0}{\alpha^3 m^2} - \frac{q E_0}{\alpha^2 m} \tag{10}$$

Solving

$$C_1 = v_0\left(1 - \frac{q B_1^2}{m\alpha}\right) \tag{11}$$

$$C_2 = \frac{qB_1}{m\alpha}\left(\frac{qB_2 E_0}{\alpha^2 m} + v_0 B_1\right) \tag{12}$$

Feeding equations 11 and 12 in 7 and 8

$$v_x = v_0\left(1 - \frac{qB_1^2}{m\alpha}\right) + \frac{qB_1}{m\alpha}\left(\frac{qB_2 E_0}{\alpha^2 m} + v_0 B_1\right)e^{\alpha t} - \frac{q^2 B_1 B_2 E_0(\alpha t + 1)}{\alpha^3 m^2} - \frac{q^2 B_1 B_2 E_0 t^2}{2\alpha m^2} \tag{13}$$

$$v_y = \left(\frac{qB_2 E_0}{\alpha^2 m} + v_0 B_1\right)\frac{e^{\alpha t}}{B_2}\left(1 - \frac{qB_1^2}{m\alpha}\right) - \frac{qE_0}{\alpha^2 m} - \frac{qE_0 t}{\alpha m} - v_0\left(1 - \frac{qB_1^2}{m\alpha}\right)\frac{B_1}{B_2} + \frac{q^2 B_1^2 E_0(\alpha t + 1)}{\alpha^3 m^2} + \frac{q^2 B_1^2 E_0 t^2}{2\alpha m^2} \tag{14}$$

Using differential definitions of velocity components in equations 13 and 14, separating the variables and integrating both sides

$$\frac{dx}{dt} = v_0\left(1 - \frac{qB_1^2}{m\alpha}\right) + \frac{qB_1}{m\alpha}\left(\frac{qB_2 E_0}{\alpha^2 m} + v_0 B_1\right)e^{\alpha t} - \frac{q^2 B_1 B_2 E_0(\alpha t + 1)}{\alpha^3 m^2} - \frac{q^2 B_1 B_2 E_0 t^2}{2\alpha m^2}$$

$$\Rightarrow \int_0^x dx = \int_0^t \left[v_0\left(1 - \frac{qB_1^2}{m\alpha}\right) + \frac{qB_1}{m\alpha}\left(\frac{qB_2 E_0}{\alpha^2 m} + v_0 B_1\right)e^{\alpha t} - \frac{q^2 B_1 B_2 E_0(\alpha t + 1)}{\alpha^3 m^2} - \frac{q^2 B_1 B_2 E_0 t^2}{2\alpha m^2}\right]dt \tag{15}$$

$$\Rightarrow x = v_0\left(1 - \frac{qB_1^2}{m\alpha}\right)t + \frac{qB_1}{m\alpha}\left(\frac{qB_2 E_0}{\alpha^2 m} + v_0 B_1\right)\frac{(e^{\alpha t} - 1)}{\alpha} - \frac{q^2 B_1 B_2 E_0}{\alpha^3 m^2}\left(\frac{\alpha t^2}{2} + t\right) - \frac{q^2 B_1 B_2 E_0 t^3}{6\alpha m^2}$$

$$\frac{dy}{dt} = \left(\frac{qB_2 E_0}{\alpha^2 m} + v_0 B_1\right)\frac{e^{\alpha t}}{B_2}\left(1 - \frac{qB_1^2}{m\alpha}\right) - \frac{qE_0}{\alpha^2 m} - \frac{qE_0 t}{\alpha m} - v_0\left(1 - \frac{qB_1^2}{m\alpha}\right)\frac{B_1}{B_2} + \frac{q^2 B_1^2 E_0(\alpha t + 1)}{\alpha^3 m^2} + \frac{q^2 B_1^2 E_0 t^2}{2\alpha m^2}$$

$$\Rightarrow \int_0^y dy = \int_0^t \left[\left(\frac{qB_2 E_0}{\alpha^2 m} + v_0 B_1\right)\frac{e^{\alpha t}}{B_2}\left(1 - \frac{qB_1^2}{m\alpha}\right) - \frac{qE_0}{\alpha^2 m} - \frac{qE_0 t}{\alpha m} - v_0\left(1 - \frac{qB_1^2}{m\alpha}\right)\frac{B_1}{B_2} + \frac{q^2 B_1^2 E_0(\alpha t + 1)}{\alpha^3 m^2} + \frac{q^2 B_1^2 E_0 t^2}{2\alpha m^2}\right]dt \tag{16}$$

$$\Rightarrow y = \left(\frac{qB_2 E_0}{\alpha^2 m} + v_0 B_1\right)\frac{(e^{\alpha t} - 1)}{\alpha B_2}\left(1 - \frac{qB_1^2}{m\alpha}\right) - \frac{qE_0 t}{\alpha^2 m} - \frac{qE_0 t^2}{2\alpha m} - v_0\left(1 - \frac{qB_1^2}{m\alpha}\right)\frac{B_1}{B_2}t + \frac{q^2 B_1^2 E_0\left(\frac{\alpha t^2}{2} + t\right)}{\alpha^3 m^2} + \frac{q^2 B_1^2 E_0 t^3}{6\alpha m^2}$$

The final position vector will be initial position vector plus displacement vector.

$$\vec{r}_2 = \vec{r}_1 + \vec{s}$$

$$\Rightarrow \vec{r}_2 = \left(0\hat{i} + 0\hat{j} + 0\hat{k}\right) + \left(x\hat{i} + y\hat{j} + z\hat{k}\right) = x\hat{i} + y\hat{j} + z\hat{k}$$

$$\Rightarrow \vec{r}_2 = \left[v_0\left(1 - \frac{qB_1^2}{m\alpha}\right)t + \frac{qB_1}{m\alpha}\left(\frac{qB_2 E_0}{\alpha^2 m} + v_0 B_1\right)\frac{(e^{\alpha t} - 1)}{\alpha} - \frac{q^2 B_1 B_2 E_0}{\alpha^3 m^2}\left(\frac{\alpha t^2}{2} + t\right) - \frac{q^2 B_1 B_2 E_0 t^3}{6\alpha m^2}\right]\hat{i} +$$

$$\left[\left(\frac{qB_2 E_0}{\alpha^2 m} + v_0 B_1\right)\frac{(e^{\alpha t} - 1)}{\alpha B_2}\left(1 - \frac{qB_1^2}{m\alpha}\right) - \frac{qE_0 t}{\alpha^2 m} - \frac{qE_0 t^2}{2\alpha m} - v_0\left(1 - \frac{qB_1^2}{m\alpha}\right)\frac{B_1}{B_2}t + \frac{q^2 B_1^2 E_0\left(\frac{\alpha t^2}{2} + t\right)}{\alpha^3 m^2} + \frac{q^2 B_1^2 E_0 t^3}{6\alpha m^2}\right]\hat{j} \tag{17}$$

Question 4

At $t=0$, a charged particle having charge q and mass m is launched from the origin with an initial velocity vector: $\vec{v}_{initial} = v_0\hat{i}$ in a region having a magnetic field $\vec{B} = B_0 e^{-ax}\hat{j}$. Find x coordinate of displacement as a function of time. Consider only electromagnetic forces.

SOLUTION

Applying Newton's second law of motion on the particle at a general time t

$$\vec{F} = m(\vec{acc})$$

$$\Rightarrow q\vec{v}\text{x}\vec{B} = m(\vec{acc}) \tag{1}$$

$$\Rightarrow q\left(v_x\hat{i}+v_y\hat{j}+v_z\hat{k}\right)\text{x}\left(B_0 e^{-ax}\hat{j}\right) = m\left(a_x\hat{i}+a_y\hat{j}+a_z\hat{k}\right)$$

Finding the cross product

$$q\left(v_x B_0 e^{-ax}\hat{k}-v_z B_0 e^{-ax}\hat{i}\right) = m\left(a_x\hat{i}+a_y\hat{j}+a_z\hat{k}\right) \tag{2}$$

Equating the vector components on left and right-hand sides of equation 2

$$a_x = -\frac{q v_z B_0 e^{-ax}}{m}$$

$$\Rightarrow \frac{dv_x}{dt} = -\frac{q v_z B_0 e^{-ax}}{m} \tag{3}$$

$$a_z = \frac{q v_x B_0 e^{-ax}}{m}$$

$$\Rightarrow \frac{dv_z}{dt} = \frac{q v_x B_0 e^{-ax}}{m} \tag{4}$$

Dividing equations 3 and 4

$$\frac{dv_x}{dv_z} = -\frac{v_z}{v_x} \tag{5}$$

Separating the variables and integrating both sides

$$\int_{v_0}^{v_x} v_x \cdot dv_x = -\int_{0}^{v_z} v_z \cdot dv_z \tag{6}$$

$$\Rightarrow v_x^2 + v_z^2 = v_0^2$$

Feeding v_z from equation 6 in equation 3

$$v_x \frac{dv_x}{dx} = -\frac{q \sqrt{v_0^2 - v_x^2}\, B_0 e^{-ax}}{m} \tag{7}$$

Separating the variables and integrating both sides

$$\int_{v_0}^{v_x} \frac{v_x \cdot dv_x}{\sqrt{v_0^2 - v_x^2}} = -\int_{0}^{x} \frac{q B_0 e^{-ax}}{m} dx \tag{8}$$

Let $v_0^2 - v_x^2 = U^2 \Rightarrow -v_x \cdot dv_x = U \cdot dU$

$$-\int_0^{\sqrt{v_0^2-v_x^2}} dU = -\int_0^x \frac{qB_0 e^{-ax}}{m}\,dx$$

$$\Rightarrow \sqrt{v_0^2 - v_x^2} = \frac{qB_0\left(1-e^{-ax}\right)}{am} \tag{9}$$

Rearranging

$$v_x = \sqrt{v_0^2 - \frac{q^2 B_0^2\left(1-e^{-ax}\right)^2}{a^2 m^2}} \tag{10}$$

Using the differential definition of velocity, separating the variables and integrating both sides

$$\int_0^x \frac{dx}{\sqrt{v_0^2 - \dfrac{q^2 B_0^2\left(1-e^{-ax}\right)^2}{a^2 m^2}}} = \int_0^t dt \tag{11}$$

Let $1-e^{-ax} = V \Rightarrow ae^{-ax}\cdot dx = dV$

$$\int_0^{1-e^{-ax}} \frac{dV}{a(1-V)\sqrt{v_0^2 - \dfrac{q^2 B_0^2 V^2}{a^2 m^2}}} = \int_0^t dt$$

$$\Rightarrow \int_0^{1-e^{-ax}} \frac{dV}{a(1-V)\dfrac{qB_0}{am}\sqrt{v_0^2 \dfrac{a^2 m^2}{q^2 B_0^2} - V^2}} = \int_0^t dt \tag{12}$$

Let $\dfrac{v_0 am}{qB_0}$ be termed as γ

$$\int_0^{1-e^{-ax}} \frac{dV}{(1-V)\sqrt{\gamma^2 - V^2}} = \frac{qB_0 t}{m} \tag{13}$$

Let $1-V = \dfrac{1}{W} \Rightarrow V = 1 - \dfrac{1}{W}$ and $dV = \dfrac{dW}{W^2}$

$$\int_{1}^{e^{ax}} \frac{dW}{W\sqrt{\gamma^2 - \left(1 - \dfrac{1}{W}\right)^2}} = \frac{qB_0 t}{m}$$

$$\Rightarrow \int_{1}^{e^{ax}} \frac{dW}{\sqrt{\left(\gamma^2 - 1\right)W^2 + 2W - 1}} = \frac{qB_0 t}{m} \tag{14}$$

Simplifying

$$\int_{1}^{e^{ax}} \frac{dW}{\sqrt{\gamma^2 - 1}\sqrt{W^2 + \dfrac{2W}{\gamma^2 - 1} - \dfrac{1}{\gamma^2 - 1}}} = \frac{qB_0 t}{m}$$

$$\Rightarrow \int_{1}^{e^{ax}} \frac{dW}{\sqrt{\gamma^2 - 1}\sqrt{W^2 + \dfrac{2W}{\gamma^2 - 1} + \left(\dfrac{1}{\gamma^2 - 1}\right)^2 - \left(\dfrac{1}{\gamma^2 - 1}\right)^2 - \dfrac{1}{\gamma^2 - 1}}} = \frac{qB_0 t}{m} \tag{15}$$

$$\int_{1}^{e^{ax}} \frac{dW}{\sqrt{\left(W + \dfrac{1}{\gamma^2 - 1}\right)^2 - \left(\dfrac{\gamma}{\gamma^2 - 1}\right)^2}} = \frac{qB_0 t\sqrt{\gamma^2 - 1}}{m}$$

$$\Rightarrow \left[\ln\left|W + \frac{1}{\gamma^2 - 1} + \sqrt{\left(W + \frac{1}{\gamma^2 - 1}\right)^2 - \left(\frac{\gamma}{\gamma^2 - 1}\right)^2}\,\right|\right]_{1}^{e^{ax}} = \frac{qB_0 t\sqrt{\gamma^2 - 1}}{m} \tag{16}$$

$$\frac{e^{ax} + \dfrac{1}{\gamma^2 - 1} + \sqrt{\left(e^{ax} + \dfrac{1}{\gamma^2 - 1}\right)^2 - \left(\dfrac{\gamma}{\gamma^2 - 1}\right)^2}}{\dfrac{\gamma^2}{\gamma^2 - 1} + \sqrt{\dfrac{\gamma^2}{\gamma^2 - 1}}} = e^{\frac{qB_0 t\sqrt{\gamma^2 - 1}}{m}} \tag{17}$$

Expanding back γ

$$\frac{e^{ax} + \dfrac{q^2 B_0^2}{v_0^2 a^2 m^2 - q^2 B_0^2} + \sqrt{\left(e^{ax} + \dfrac{q^2 B_0^2}{v_0^2 a^2 m^2 - q^2 B_0^2}\right)^2 - \left(\dfrac{qB_0 v_0 am}{v_0^2 a^2 m^2 - q^2 B_0^2}\right)^2}}{\dfrac{v_0^2 a^2 m^2}{v_0^2 a^2 m^2 - q^2 B_0^2} + \sqrt{\dfrac{v_0^2 a^2 m^2}{v_0^2 a^2 m^2 - q^2 B_0^2}}} = e^{\frac{t\sqrt{v_0^2 a^2 m^2 - q^2 B_0^2}}{m}} \tag{18}$$

Question 5

❖ ❖ ❖

A particle of mass m and charge q is launched with a velocity v_0 along a horizontal rough surface. The friction coefficient for the particle surface pair varies as $\mu = \dfrac{\mu_0}{(a+bX)^2}$ where X is the distance measured from the launching point along the direction of v_0 and μ_0, a, b are positive constants. There exists a uniform magnetic field B out of the plane of paper in the region. Find the time taken by the particle to fall off the surface. Ignore gravitational, buoyant and air resistance forces.

SOLUTION

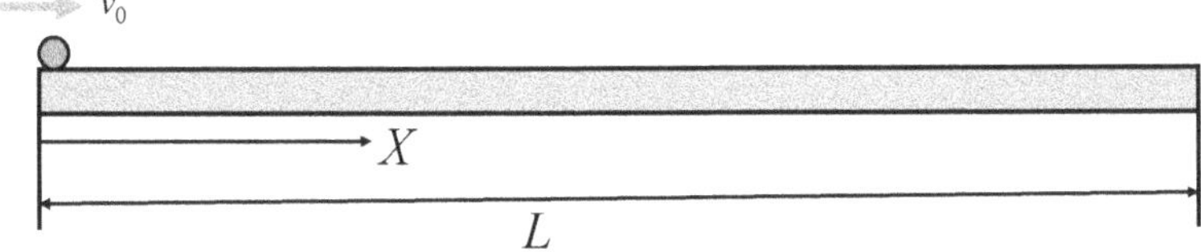

Applying Newton's second law of motion on the particle at a general time t

$$-\mu N = m(acc) \tag{1}$$

$$N = qvB \tag{2}$$

Feeding equation 2 in 1 and using the given expression of friction coefficient

$$-\frac{\mu_0 qvB}{m(a+bx)^2} = \frac{v \cdot dv}{dx} \tag{3}$$

Separating the variables and integrating both sides

$$\int_{v_0}^{v} dv = -\int_{0}^{x} \frac{\mu_0 qB}{m(a+bx)^2} dx$$

$$\Rightarrow v - v_0 = \frac{\mu_0 qB}{mb}\left[\frac{1}{a+bx} - \frac{1}{a}\right] \tag{4}$$

Rearranging

$$v = v_0 - \frac{\mu_0 qBx}{ma(a+bx)} \tag{5}$$

Using the differential definition of velocity, separating the variables and integrating both sides

$$\frac{dx}{dt} = v_0 - \frac{\mu_0 qBx}{ma(a+bx)}$$

$$\Rightarrow \int_{0}^{x} \frac{dx}{v_0 - \dfrac{\mu_0 qBx}{ma(a+bx)}} = \int_{0}^{t} dt \tag{6}$$

Simplifying

$$\int_{0}^{x} \frac{(a+bx)dx}{(a+bx)v_0 - cx} = \int_{0}^{t} dt$$

$$\Rightarrow \int_{0}^{x} \frac{(a+bx)dx}{av_0 + (bv_0 - c)x} = \int_{0}^{t} dt \tag{7}$$

$$a\int_0^x \frac{dx}{av_0 + (bv_0 - c)x} + b\int_0^x \frac{x \cdot dx}{av_0 + (bv_0 - c)x} = t$$

$$\Rightarrow a\int_0^x \frac{dx}{av_0 + (bv_0 - c)x} + \frac{b}{(bv_0 - c)}\int_0^x \frac{(bv_0 - c)x \cdot dx}{av_0 + (bv_0 - c)x} = t \tag{8}$$

$$a\int_0^x \frac{dx}{av_0 + (bv_0 - c)x} + \frac{b}{(bv_0 - c)}\int_0^x \frac{\left[av_0 + (bv_0 - c)x - av_0\right] \cdot dx}{av_0 + (bv_0 - c)x} = t$$

$$\Rightarrow a\int_0^x \frac{dx}{av_0 + (bv_0 - c)x} + \frac{b}{(bv_0 - c)}\int_0^x dx - \frac{abv_0}{(bv_0 - c)}\int_0^x \frac{dx}{av_0 + (bv_0 - c)x} = t \tag{9}$$

$$\frac{bx}{bv_0 - c} - \frac{ac}{(bv_0 - c)^2}\ln\left|\frac{av_0 + (bv_0 - c)x}{av_0}\right| = t \tag{10}$$

The time at which the particle leaves the surface is the t at which $x=L$

$$t_{surface\ departure} = \frac{bL}{bv_0 - c} - \frac{ac}{(bv_0 - c)^2}\ln\left|\frac{av_0 + (bv_0 - c)L}{av_0}\right| \tag{11}$$

Expanding back c

$$t_{surface\ departure} = \frac{bL}{bv_0 - \dfrac{\mu_0 qB}{ma}} - \frac{\dfrac{\mu_0 qB}{m}}{\left(bv_0 - \dfrac{\mu_0 qB}{ma}\right)^2}\ln\left|\frac{av_0 + \left(bv_0 - \dfrac{\mu_0 qB}{ma}\right)L}{av_0}\right| \tag{12}$$

Question 6

❖ ❖ ❖

A particle of mass m and charge q is launched with a horizontal velocity v_0 from a location shown. A large sheet carrying uniformly distributed (charge per unit area $= \sigma$)is present at a large distance to the right as shown. The permittivity of the medium however, is variable. It varies with the vertical distance Y from the ground as: $\varepsilon = a - bY$. Find the horizontal separation of the location where the particle strikes the ground from the launching point. The acceleration due to gravity on the Earth's surface is g. Air resistance and buoyancy are ignored

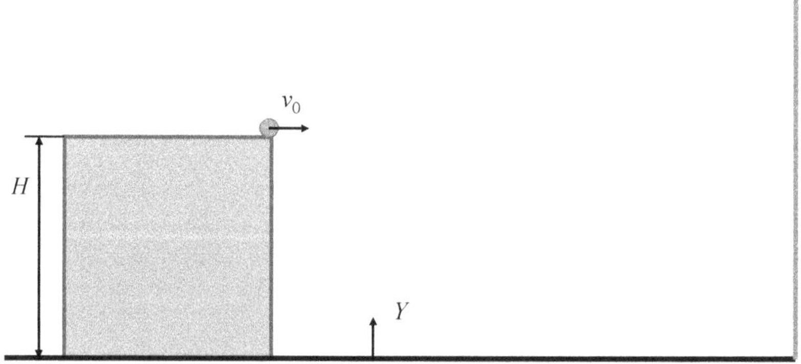

SOLUTION

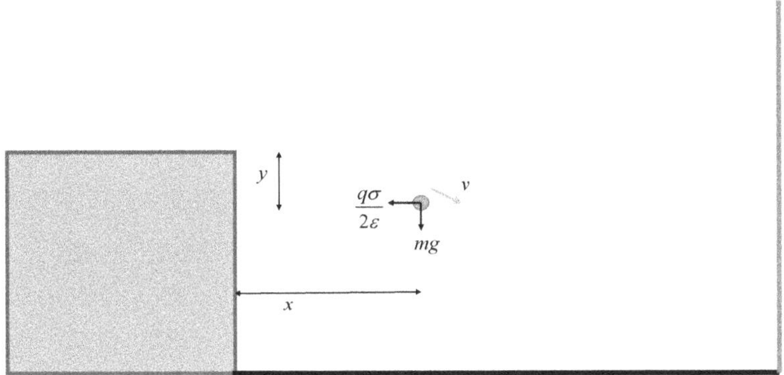

Applying Newton's second law of motion on the particle at a general time t

$$-\frac{q\sigma}{2\varepsilon} = m\left(acc_x\right)$$

$$\Rightarrow \frac{dv_x}{dt} = -\frac{q\sigma}{2m\left[a - b\left(H - y\right)\right]} \tag{1}$$

$$mg = m\left(acc_y\right)$$

$$\Rightarrow \frac{dv_y}{dt} = g \tag{2}$$

Here, the well-known formula for the electric field of an infinite sheet: $\dfrac{\sigma}{2\varepsilon}$ has been used. Separating the variables and integrating both sides in equation 2

$$\int_{0}^{v_y} dv_y = g\int_{0}^{t} dt \tag{3}$$

$$\Rightarrow v_y = gt$$

Using the differential definition of velocity, separating the variables and integrating both sides in equation 3

$$\frac{dy}{dt} = gt$$

$$\Rightarrow \int_0^y dy = g \int_0^t t \cdot dt \tag{4}$$

$$\Rightarrow y = \frac{gt^2}{2}$$

Feeding equation 4 in 1

$$\frac{dv_x}{dt} = -\frac{q\sigma}{2m\left(a - bH + \dfrac{bgt^2}{2}\right)} \tag{5}$$

Separating the variables and integrating both sides

$$\int_{v_0}^{v_x} dv_x = -\int_0^t \frac{q\sigma}{2m\left(a - bH + \dfrac{bgt^2}{2}\right)} dt$$

$$\Rightarrow v_x - v_0 = -\frac{q\sigma}{2m} \int_0^t \frac{dt}{a - bH + \dfrac{bgt^2}{2}} \tag{6}$$

Simplifying

$$v_x - v_0 = -\frac{q\sigma}{2m\dfrac{bg}{2}} \int_0^t \frac{dt}{t^2 + \left(\sqrt{\dfrac{2(a - bH)}{bg}}\right)^2}$$

$$\Rightarrow v_x = v_0 - \frac{q\sigma}{mbg} \frac{1}{\sqrt{\dfrac{2(a - bH)}{bg}}} \tan^{-1}\left(\frac{t}{\sqrt{\dfrac{2(a - bH)}{bg}}}\right) \tag{7}$$

Using the differential definition of velocity, separating the variables and integrating both sides in equation 7

$$\frac{dx}{dt} = v_0 - \frac{q\sigma}{mbg}\frac{1}{\sqrt{\frac{2(a-bH)}{bg}}}\tan^{-1}\left(\frac{t}{\sqrt{\frac{2(a-bH)}{bg}}}\right)$$

$$\Rightarrow \int_0^x dx = v_0 \int_0^t dt - \int_0^t \frac{q\sigma}{m\sqrt{2bg(a-bH)}}\tan^{-1}\left(t\sqrt{\frac{bg}{2(a-bH)}}\right)dt \tag{8}$$

$$\Rightarrow x = v_0 t - \frac{q\sigma}{m\sqrt{2bg(a-bH)}}\int_0^t \tan^{-1}\left(t\sqrt{\frac{bg}{2(a-bH)}}\right)dt$$

$$\text{Let } \tan^{-1}\left(t\sqrt{\frac{bg}{2(a-bH)}}\right) = p \Rightarrow t\sqrt{\frac{bg}{2(a-bH)}} = \tan p \Rightarrow dt = \sqrt{\frac{2(a-bH)}{bg}}\sec^2 p \cdot dp$$

$$x = v_0 t - \frac{q\sigma}{m\sqrt{2bg(a-bH)}}\int_0^{\tan^{-1}\left(t\sqrt{\frac{bg}{2(a-bH)}}\right)} p\sqrt{\frac{2(a-bH)}{bg}}\sec^2 p \cdot dp \tag{9}$$

Using integration by parts

$$x = v_0 t - \frac{q\sigma}{m\sqrt{2bg(a-bH)}}\sqrt{\frac{2(a-bH)}{bg}}\left[p\tan p - \int \tan p \cdot dp\right]_0^{\tan^{-1}\left(t\sqrt{\frac{bg}{2(a-bH)}}\right)}$$

$$\Rightarrow x = v_0 t - \frac{q\sigma}{m\sqrt{2bg(a-bH)}}\sqrt{\frac{2(a-bH)}{bg}}\left[p\tan p - \ln|\sec p|\right]_0^{\tan^{-1}\left(t\sqrt{\frac{bg}{2(a-bH)}}\right)} \tag{10}$$

$$x = v_0 t - \frac{q\sigma}{mbg}\left[t\sqrt{\frac{bg}{2(a-bH)}}\tan^{-1}\left(t\sqrt{\frac{bg}{2(a-bH)}}\right) - \ln\left|\sec\left\{\tan^{-1}\left(t\sqrt{\frac{bg}{2(a-bH)}}\right)\right\}\right|\right] \tag{11}$$

The particle touches the ground when $y=H$. Using equation 4, it can be seen that this shall occur at $t = \sqrt{\dfrac{2H}{g}}$. Feeding this value of t in equation 11 gives the horizontal displacement when the particle strikes the ground.

Question 7

❖ ❖ ❖

Aparticle with charge q and mass m rests on the top of an inverted hemisphere. Another particle of charge $-Q$ is fixed at the shown location. The movable particle is given a gentle push. Find the location where it breaks contact with the hemisphere. Gravitational, air resistance and buoyant forces are neglected force and the hemisphere is fixed.

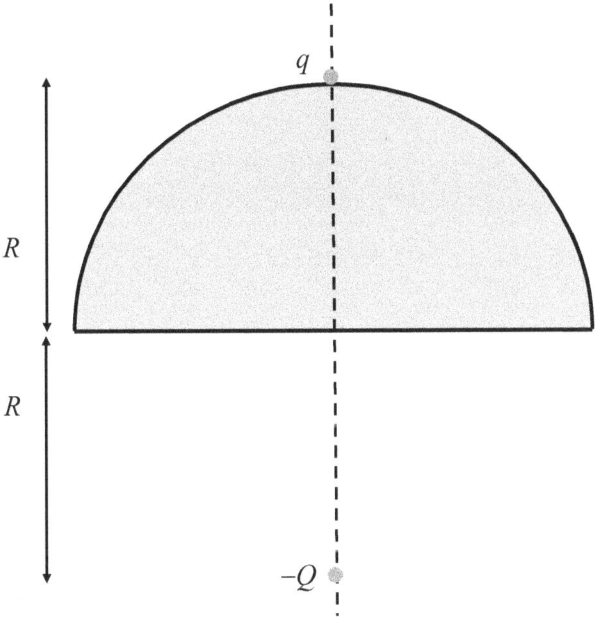

SOLUTION

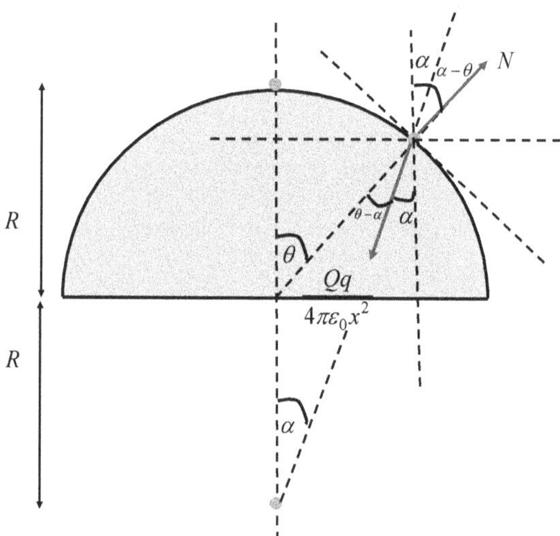

Applying Newton's second law of motion on the particle in radial and tangential directions, at a general time t

$$\frac{Qq\cos(\theta-\alpha)}{4\pi\varepsilon_0 x^2} - N = \frac{mv^2}{R} \tag{1}$$

$$\frac{Qq\sin(\theta-\alpha)}{4\pi\varepsilon_0 x^2} = \frac{mv\cdot dv}{ds}$$

$$\Rightarrow \frac{Qq\sin(\theta-\alpha)}{4\pi\varepsilon_0 x^2} = \frac{mv\cdot dv}{R\cdot d\theta} \tag{2}$$

The variables α and x need to be expressed in terms of θ. This can be done using geometry as shown below

$$\tan\alpha = \frac{R\sin\theta}{R + R\cos\theta} = \frac{2\sin\dfrac{\theta}{2}\cos\dfrac{\theta}{2}}{2\cos^2\dfrac{\theta}{2}} = \tan\frac{\theta}{2} \tag{3}$$

$$\Rightarrow \alpha = \frac{\theta}{2}$$

$$x = \sqrt{R^2 \sin^2\theta + \left(R\cos\theta + R\right)^2}$$

$$\Rightarrow x = \sqrt{2R^2\left(1 + \cos\theta\right)}$$

$$\Rightarrow x = \sqrt{4R^2\cos^2\frac{\theta}{2}} \tag{4}$$

$$\Rightarrow x = 2R\cos\frac{\theta}{2}$$

Feeding equations 3 and 4 in equation 2

$$\frac{v \cdot dv}{d\theta} = \frac{Qq\sin\dfrac{\theta}{2}}{16\pi\varepsilon_0 mR\cos^2\dfrac{\theta}{2}} \tag{5}$$

Separating the variables and integrating both sides

$$\int_0^v v \cdot dv = \frac{Qq}{16\pi\varepsilon_0 mR}\int_0^\theta \frac{\sin\dfrac{\theta}{2}}{\cos^2\dfrac{\theta}{2}}\,d\theta \tag{6}$$

Let $\cos\dfrac{\theta}{2} = U \Rightarrow -\dfrac{1}{2}\sin\dfrac{\theta}{2} \cdot d\theta = dU$

$$\frac{v^2}{2} = -\frac{Qq}{16\pi\varepsilon_0 mR}\int_1^{\cos\frac{\theta}{2}} \frac{2 \cdot dU}{U^2}$$

$$\Rightarrow v^2 = \frac{Qq}{4\pi\varepsilon_0 mR}\left(\frac{1}{\cos\dfrac{\theta}{2}} - 1\right) \tag{7}$$

Feeding equation 3,4,7 in 1 gives an expression for Normal reaction as a function of location.

$$\frac{Qq}{16\pi\varepsilon_0 R^2 \cos\dfrac{\theta}{2}} - N = \frac{Qq}{4\pi\varepsilon_0 R^2}\left(\frac{1}{\cos\dfrac{\theta}{2}} - 1\right)$$

$$\Rightarrow N = \frac{Qq}{4\pi\varepsilon_0 R^2}\left[\frac{1}{4\cos\dfrac{\theta}{2}} - \frac{1}{\cos\dfrac{\theta}{2}} + 1\right] \tag{8}$$

$$\Rightarrow N = \frac{Qq}{4\pi\varepsilon_0 R^2}\left[1 - \frac{3}{4\cos\dfrac{\theta}{2}}\right]$$

Contact loss occurs where $N=0$.

$$\cos\frac{\theta}{2} = \frac{3}{4}$$

$$\Rightarrow \frac{\theta}{2} = \cos^{-1}\frac{3}{4} \tag{9}$$

$$\Rightarrow \theta = 2\cos^{-1}\frac{3}{4}$$

Question 8

A particle with charge q and mass m rests at the bottommost position of a vertical semicircular track of radius R. Another particle of charge $-Q$ is fixed at the shown location. The former particle is imparted a horizontal velocity v_0. Find the minimum value of v_0 for which this particle completes a vertical circle. The permittivity of free space is ε_0. Gravitational, air resistance and bouyant forces are neglected and the track is fixed.

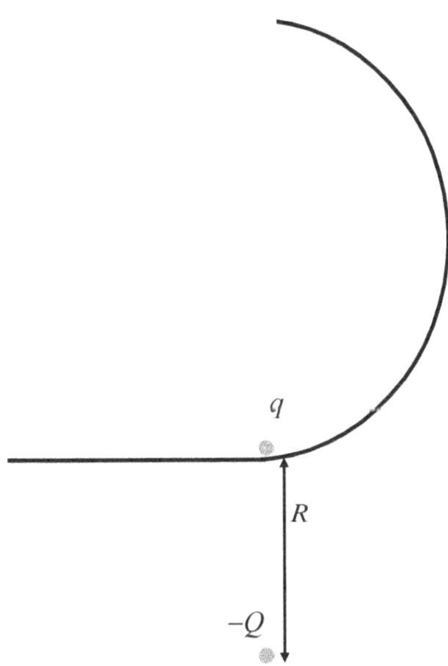

SOLUTION

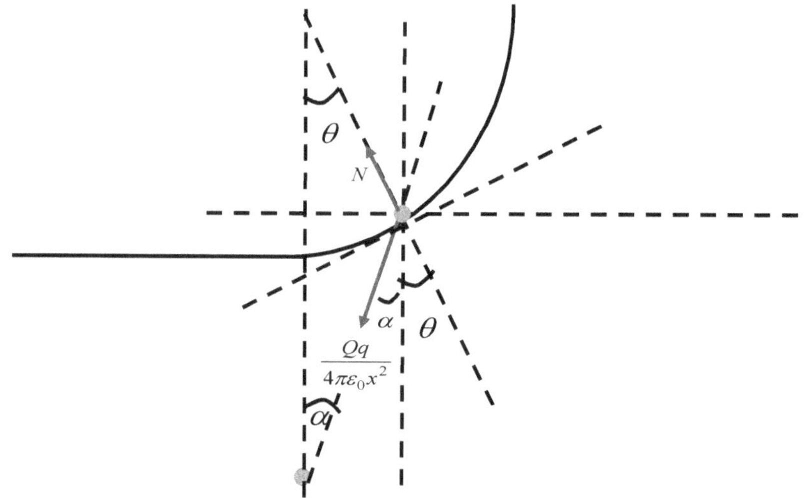

Applying Newton's second law of motion on the particle in radial and tangential directions, at a general time t

$$N - \frac{Qq\cos(\theta+\alpha)}{4\pi\varepsilon_0 x^2} = \frac{mv^2}{R} \tag{1}$$

$$-\frac{Qq\sin(\theta+\alpha)}{4\pi\varepsilon_0 x^2} = \frac{mv \cdot dv}{ds}$$

$$\Rightarrow -\frac{Qq\sin(\theta+\alpha)}{4\pi\varepsilon_0 x^2} = \frac{mv \cdot dv}{R \cdot d\theta} \tag{2}$$

Equations 1 and 2 can be written as

$$N - \frac{Qq(\cos\alpha\cos\theta - \sin\alpha\sin\theta)}{4\pi\varepsilon_0 x^2} = \frac{mv^2}{R} \tag{3}$$

$$-\frac{Qq(\sin\theta\cos\alpha + \cos\theta\sin\alpha)}{4\pi\varepsilon_0 x^2} = \frac{mv \cdot dv}{R \cdot d\theta} \tag{4}$$

The variables α and x need to be expressed in terms of θ. This can be done using geometry as shown below

$$x = \sqrt{(R\sin\theta)^2 + (R+R-R\cos\theta)^2}$$
$$\Rightarrow x = \sqrt{R^2\left[\sin^2\theta + (2-\cos\theta)^2\right]}$$
$$\Rightarrow x = \sqrt{R^2(5-4\cos\theta)}$$
$$\Rightarrow x = R\sqrt{5-4\cos\theta}$$
$$\tag{5}$$

$$\cos\alpha = \frac{R+R-R\cos\theta}{x}$$
$$\Rightarrow \cos\alpha = \frac{R+R-R\cos\theta}{R\sqrt{5-4\cos\theta}}$$
$$\Rightarrow \cos\alpha = \frac{2-\cos\theta}{\sqrt{5-4\cos\theta}}$$
$$\tag{6}$$

$$\sin\alpha = \frac{R\sin\theta}{x}$$
$$\Rightarrow \sin\alpha = \frac{\sin\theta}{\sqrt{5-4\cos\theta}}$$
$$\tag{7}$$

Feeding equations 5,6,7 in 3,4

$$N - \frac{Qq(2\cos\theta - 1)}{4\pi\varepsilon_0 R^2 (5-4\cos\theta)^{1.5}} = \frac{mv^2}{R}$$
$$\tag{8}$$

$$-\frac{Qq\sin\theta}{2\pi\varepsilon_0 mR(5-4\cos\theta)^{1.5}} = \frac{v\cdot dv}{d\theta}$$
$$\tag{9}$$

Separating the variables and integrating both sides in equation 9

$$\int_{v_0}^{v} v\cdot dv = -\int_{0}^{\theta} \frac{Qq\sin\theta\cdot d\theta}{2\pi\varepsilon_0 mR(5-4\cos\theta)^{1.5}}$$
$$\tag{10}$$

Let $\cos\theta = U \Rightarrow -\sin\theta\cdot d\theta = dU$

$$\frac{v^2 - v_0^2}{2} = \frac{Qq}{2\pi\varepsilon_0 mR} \int_1^{\cos\theta} \frac{dU}{\left(5 - 4U\right)^{1.5}} \tag{11}$$

$$\frac{v^2 - v_0^2}{2} = \frac{Qq}{2\pi\varepsilon_0 mR} \left[\frac{\left(5 - 4U\right)^{-0.5}}{\left(-0.5\right)\left(-4\right)} \right]_1^{\cos\theta}$$

$$\Rightarrow v^2 = v_0^2 + \frac{Qq}{2\pi\varepsilon_0 mR} \left[\frac{1}{\sqrt{5 - 4\cos\theta}} - 1 \right] \tag{12}$$

Feeding equation 12 in 8 gives an expression for normal reaction as a function of location

$$N = \frac{Qq\left(2\cos\theta - 1\right)}{4\pi\varepsilon_0 R^2 \left(5 - 4\cos\theta\right)^{1.5}} + \frac{m}{R} v_0^2 + \frac{Qq}{2\pi\varepsilon_0 R^2} \left[\frac{1}{\sqrt{5 - 4\cos\theta}} - 1 \right] \tag{13}$$

The condition for completion of a vertical circle **is** $N \geq 0$ at $\theta = \pi$

$$\frac{-Qq}{36\pi\varepsilon_0 R^2} + \frac{mv_0^2}{R} - \frac{Qq}{3\pi\varepsilon_0 R^2} \geq 0 \tag{14}$$

$$v_0 \geq \sqrt{\frac{13Qq}{36\pi\varepsilon_0 mR}}$$

$$\Rightarrow \left(v_0\right)_{min} = \sqrt{\frac{13Qq}{36\pi\varepsilon_0 mR}} \tag{15}$$

Question 9

Aparticle of mass m and charge q is held at the base of an inclined plane. Another point charge $-Q$ is fixed at the location shown. The incline surface is rough with the friction coefficient for the former particle and surface pair being μ. At $t=0$, the former charge is released. Find the velocity with which it leaves the incline. Ignore gravitational, air resistance and buoyant forces. The two charges lie in the same plane. Assume that the electrostatic force acting on the charge at $t=0$ is enough to cause it to begin traversing the inclined plane. The permittivity of free space is ε_0.

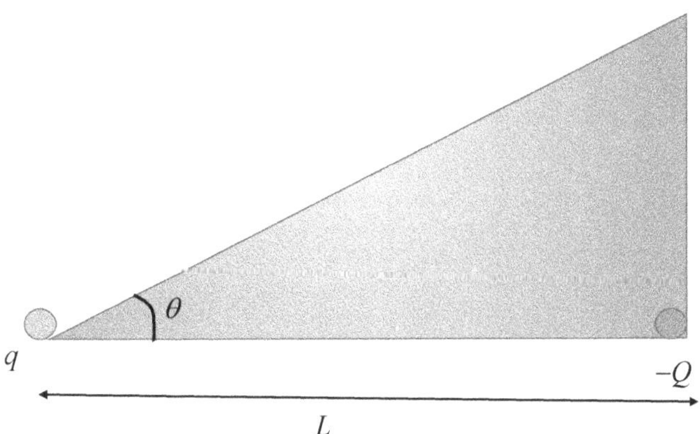

SOLUTION

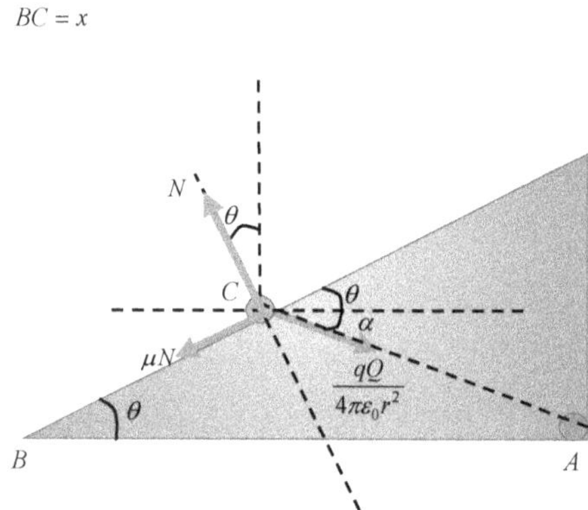

Applying Newton's second law of motion on the particle at a general time *t*

$$\frac{qQ\cos(\alpha+\theta)}{4\pi\varepsilon r^2} - \mu N = m(acc) \tag{1}$$

$$\frac{qQ}{4\pi\varepsilon r^2}\sin(\alpha+\theta) = N \tag{2}$$

Feeding *N* from equation 2 in 1

$$v\frac{dv}{dx} = \frac{qQ\left[\cos(\alpha+\theta)-\mu\sin(\alpha+\theta)\right]}{4\pi\varepsilon m r^2} \tag{3}$$

The variables α and r need to be expressed in terms of x. This can be done using geometry as shown below

$$r = \sqrt{x^2\sin^2\theta+\left(L-x\cos\theta\right)^2} \tag{4}$$

$$\cos\alpha = \frac{L-x\cos\theta}{r} \tag{5}$$

$$\sin\alpha = \frac{x\sin\theta}{r} \tag{6}$$

Feeding equations 4, 5, 6 in equation 3

$$\frac{vdv}{dx} = \frac{qQ}{4\pi\varepsilon mr^2}\left[\frac{L\cos\theta - x\cos^2\theta}{r} - \frac{x\sin^2\theta}{r} - \frac{\mu x\sin\theta\cos\theta}{r} - \frac{\mu L\sin\theta}{r} + \frac{\mu x\sin\theta\cos\theta}{r}\right]$$

$$\Rightarrow \frac{vdv}{dx} = \frac{qQ\left[L(\cos\theta - \mu\sin\theta) - x\right]}{4\pi\varepsilon m\left[x^2 + L^2 - 2Lx\cos\theta\right]^{1.5}} \tag{7}$$

Separating the variables and integrating both sides

$$\int_0^v vdv = \frac{qQ}{4\pi\varepsilon m}\int_0^x \frac{L(\cos\theta - \mu\sin\theta) - x}{\left[x^2 + L^2 - 2Lx\cos\theta\right]^{1.5}}dx \tag{8}$$

Let $L\cos\theta - x = p \Rightarrow -dx = dp$. Also, $L^2\cos^2\theta + x^2 - 2Lx\cos\theta = p^2 \Rightarrow x^2 - 2Lx\cos\theta = p^2 - L^2\cos^2\theta$

$$\int_0^v vdv = -\frac{qQ}{4\pi\varepsilon m}\int_{L\cos\theta}^{L\cos\theta - x} \frac{p - \mu L\sin\theta}{\left[p^2 + L^2\sin^2\theta\right]^{\frac{3}{2}}}dp \tag{9}$$

Let $L^2\sin^2\theta + p^2 = V^2 \Rightarrow pdp = VdV$

$$\frac{v^2}{2} = \frac{qQ\mu L\sin\theta}{4\pi\varepsilon m}\int_L^{\sqrt{L^2+x^2-2Lx\cos\theta}} \frac{dV}{V^2\sqrt{V^2 - L^2\sin^2\theta}} - \frac{qQ}{4\pi\varepsilon m}\int_L^{\sqrt{L^2+x^2-2Lx\cos\theta}} \frac{dV}{V^2} \tag{10}$$

$$\frac{v^2}{2} = \frac{qQ\mu L\sin\theta}{4\pi\varepsilon m}\int_L^{\sqrt{L^2+x^2-2Lx\cos\theta}} \frac{dV}{V^2\sqrt{V^2 - L^2\sin^2\theta}} + \frac{qQ}{4\pi\varepsilon m}\left[\frac{1}{\sqrt{L^2 + x^2 - 2Lx\cos\theta}} - \frac{1}{L}\right] \tag{11}$$

Let $V = \frac{1}{U} \Rightarrow dV = -\frac{dU}{U^2}$

$$\frac{v^2}{2} = -\frac{qQ\mu L\sin\theta}{4\pi\varepsilon m}\int_{\frac{1}{L}}^{\frac{1}{\sqrt{L^2+x^2-2Lx\cos\theta}}} \frac{U\cdot dU}{\sqrt{1 - U^2 L^2\sin^2\theta}} + \frac{qQ}{4\pi\varepsilon m}\left[\frac{1}{\sqrt{L^2 + x^2 - 2Lx\cos\theta}} - \frac{1}{L}\right] \tag{12}$$

Let $1 - U^2 L^2\sin^2\theta = K^2 \Rightarrow -\left(L^2\sin^2\theta\right)U\cdot dU = K\cdot dK$

$$\frac{v^2}{2} = \frac{qQ\mu}{4\pi\varepsilon_0 mL\sin\theta} \int\limits_{\cos\theta}^{\sqrt{\frac{L^2\cos^2\theta + x^2 - 2Lx\cos\theta}{L^2 + x^2 - 2Lx\cos\theta}}} dK + \frac{qQ}{4\pi\varepsilon_0 m}\left[\frac{1}{\sqrt{L^2 + x^2 - 2Lx\cos\theta}} - \frac{1}{L}\right] \tag{13}$$

$$v = \sqrt{\frac{qQ}{2\pi\varepsilon_0 m}\left[\frac{\mu}{L\sin\theta}\sqrt{\frac{L^2\cos^2\theta + x^2 - 2Lx\cos\theta}{L^2 + x^2 - 2Lx\cos\theta}} - \frac{\mu\cos\theta}{L\sin\theta} + \frac{1}{\sqrt{L^2 + x^2 - 2Lx\cos\theta}} - \frac{1}{L}\right]} \tag{14}$$

The exit velocity is the value of v at $x = L\sec\theta$

$$v_{exit} = \sqrt{\frac{qQ}{2\pi\varepsilon_0 m}\left[\frac{\mu}{L\sin\theta}\sqrt{\frac{L^2\cos^2\theta + L^2\sec^2\theta - 2L^2}{L^2 + L^2\sec^2\theta - 2L^2}} - \frac{\mu\cos\theta}{L\sin\theta} + \frac{1}{\sqrt{L^2 + L^2\sec^2\theta - 2L^2}} - \frac{1}{L}\right]}$$

$$\Rightarrow v_{exit} = \sqrt{\frac{qQ}{2\pi\varepsilon_0 mL}\left[\frac{\mu\cot\theta}{\sin\theta}\sqrt{\cos^2\theta + \sec^2\theta - 2} - \mu\cot\theta + \cot\theta - 1\right]} \tag{15}$$

Question 10

A particle of mass m and charge q is released from the top most location of a surface as shown. A uniform electric field E exists in the region in the negative X direction. Gravitational, air resistance and buoyant forces are neglected. Find the function f for which the particle reaches the bottom of the surface in the shortest possible time.

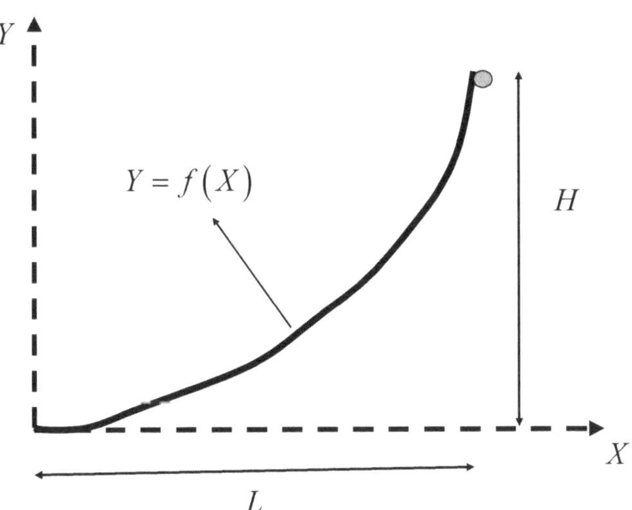

SOLUTION

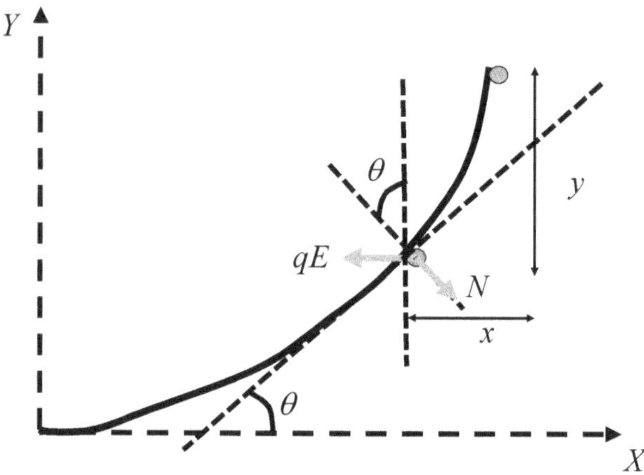

Applying Newton's law second of motion on the particle at a general time t

$$qE - N\sin\theta = ma_x \tag{1}$$

$$N\cos\theta = ma_y \tag{2}$$

Eliminating N from equations 1 and 2

$$qE - ma_y\tan\theta = ma_x \tag{3}$$

Since the particle slides on the curve, its coordinates at any instant will satisfy the curve equation

$$H - y = f(L - x) \tag{4}$$

Differentiating both sides with respect to time

$$v_y = f'(L - x)v_x \tag{5}$$

Differentiating both sides with respect to time

$$a_y = f'(L-x)a_x - f''(L-x)v_x^2 \tag{6}$$

Feeding equation 6 in equation 3

$$\frac{qE}{m} = a_x + f'(L-x)a_x \tan\theta - f''(L-x)v_x^2 \tan\theta \tag{7}$$

Further,

$$\tan\theta = \frac{dY}{dX} = \frac{d\{f(L-x)\}}{d(L-x)} = f'(L-x) \tag{8}$$

Feeding equation 8 in equation 7

$$\frac{v_x dv_x}{dx} = \frac{qE}{m\left[1+\{f'(L-x)\}^2\right]} + \frac{f''(L-x)f'(L-x)v_x^2}{1+\{f'(L-x)\}^2} \tag{9}$$

Let $v_x^2 = U \Rightarrow 2v_x dv_x = dU$

$$\frac{dU}{dx} - \left[\frac{2f''(L-x)f'(L-x)}{1+\{f'(L-x)\}^2}\right]U = \frac{2qE}{m\left[1+\{f'(L-x)\}^2\right]} \tag{10}$$

Equation 10 is a first order linear ordinary differential equation in $U(x)$ whose general solution can be given by the method of integrating factor

$$U(I.F) = \int \frac{2qE(I.F)dx}{m\left[1+\{f'(L-x)\}^2\right]} + C_0 \tag{11}$$

Where C_0 is an arbitrary integration constant and $I.F$ is integrating factor given by

$$I.F = e^{-\int\left[\frac{2f''(L-x)f'(L-x)}{1+\{f'(L-x)\}^2}\right]dx} = e^{\ln\left|1+\{f'(L-x)\}^2\right|} = 1+\{f'(L-x)\}^2 \tag{12}$$

Feeding equation 12 in equation 11

$$v_x^2 = \frac{\dfrac{2qEx}{m} + C_0}{1 + \left\{ f'(L-x) \right\}^2} \tag{13}$$

The constant C_0 is to be found by the initial condition: $v_x(x=0)=0$

$$C_0 = 0 \tag{14}$$

Feeding equation 14 in equation 13

$$v_x = \sqrt{\frac{2qEx}{m\left[1 + \left\{ f'(L-x) \right\}^2\right]}} \tag{15}$$

Using the differential definition of velocity, separating the variables and integrating both sides

$$\int_0^x \sqrt{\frac{1 + \left\{ f'(L-x) \right\}^2}{x}}\, dx = \sqrt{\frac{2qE}{m}} \int_0^t dt \tag{16}$$

Using equation 16, it can be seen that time taken to reach the bottommost point is

$$t_0 = \sqrt{\frac{m}{2qE}} \int_0^L \sqrt{\frac{1 + \left\{ f'(L-x) \right\}^2}{x}}\, dx \tag{17}$$

As per the Euler Lagrange theorem [from calculus of variations], stationary values of t_0 are given by

$$\frac{\partial F}{\partial f} = \frac{d}{dx}\left(\frac{\partial F}{\partial f'} \right) \tag{18}$$

Where $F = \sqrt{\dfrac{1 + \left\{ f'(L-x) \right\}^2}{x}}$

$$\frac{d}{dx}\left(\frac{1}{2\sqrt{\dfrac{1+\left\{f'(L-x)\right\}^2}{x}}}\,\frac{2f'(L-x)}{x}\right)=0 \tag{19}$$

$$\Rightarrow \frac{f'(L-x)}{\sqrt{x}\sqrt{1+\left\{f'(L-x)\right\}^2}}=C_1$$

Rearranging

$$f'(L-x)=\frac{C_1\sqrt{x}}{\sqrt{1-C_1^2 x}} \tag{20}$$

Replacing x by L-x

$$f'(x)=\frac{C_1\sqrt{L-x}}{\sqrt{1-C_1^2(L-x)}} \tag{21}$$

Separating the variables and integrating both sides

$$\int df = \int \frac{C_1\sqrt{L-x}}{\sqrt{1-C_1^2(L-x)}}\,dx \tag{22}$$

Let $L-x=U^2 \Rightarrow dx=-2U\cdot dU$

$$f(x)=-2C_1\int \frac{U^2\cdot dU}{\sqrt{1-C_1^2 U^2}} \tag{23}$$

Simplifying

$$f(x)=\frac{2}{C_1}\int \frac{-C_1^2 U^2}{\sqrt{1-C_1^2 U^2}}\,dU \tag{24}$$

$$\Rightarrow f(x)=\frac{2}{C_1}\int \frac{1-C_1^2 U^2 -1}{\sqrt{1-C_1^2 U^2}}\,dU$$

$$f(x) = \frac{2}{C_1} \int \sqrt{1 - C_1^2 U^2} \, dU - \frac{2}{C_1} \int \frac{dU}{\sqrt{1 - C_1^2 U^2}}$$

$$\Rightarrow f(x) = 2 \int \sqrt{\frac{1}{C_1^2} - U^2} \, dU - \frac{2}{C_1^2} \int \frac{dU}{\sqrt{\frac{1}{C_1^2} - U^2}} \tag{25}$$

Using the standard results:
$$\int \frac{dZ}{\sqrt{S^2 - Z^2}} = \sin^{-1} \frac{Z}{S} \quad \text{and}$$

$$\int \sqrt{S^2 - Z^2} \, dZ = \frac{Z\sqrt{S^2 - Z^2}}{2} + \frac{1}{2} S^2 \sin^{-1} \frac{Z}{S}$$

$$f(x) = U\sqrt{\frac{1}{C_1^2} - U^2} + \frac{1}{C_1^2} \sin^{-1}(UC_1) - \frac{2}{C_1^2} \sin^{-1}(UC_1) + C_2$$

$$\Rightarrow f(x) = U\sqrt{\frac{1}{C_1^2} - U^2} - \frac{1}{C_1^2} \sin^{-1}(UC_1) + C_2 \tag{26}$$

Expanding back U

$$f(x) = \sqrt{L - x}\sqrt{\frac{1}{C_1^2} - L + x} - \frac{1}{C_1^2} \sin^{-1}\left(C_1\sqrt{L - x}\right) + C_2 \tag{27}$$

The constants C_1 and C_2 need to be found using the boundary conditions: $f(0) = 0$ and $f(L) = H$. This evaluation is however, not possible analytically and will have to be done numerically.

Question 11

❖ ❖ ❖

A particle of mass m and charge q is released from rest at a distance H from a hollow hemisphere, along its axis as shown. The hemisphere carries a charge Q distributed uniformly over its surface. Find the velocity with which the particle enters the hemisphere. Neglect gravitational, air resistance and buoyant forces. The permittivity of free space is ε_0. The hemisphere is fixed.

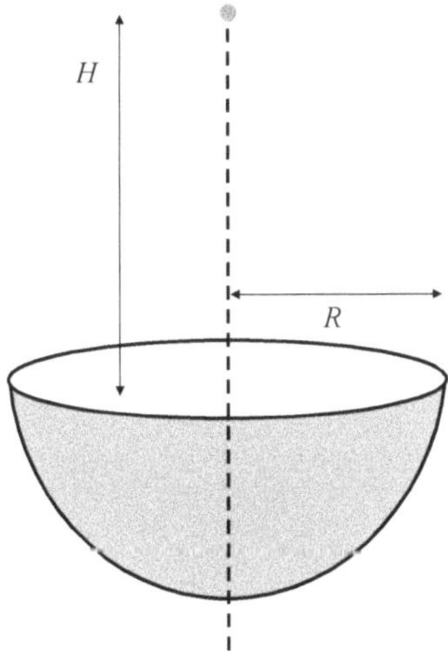

SOLUTION

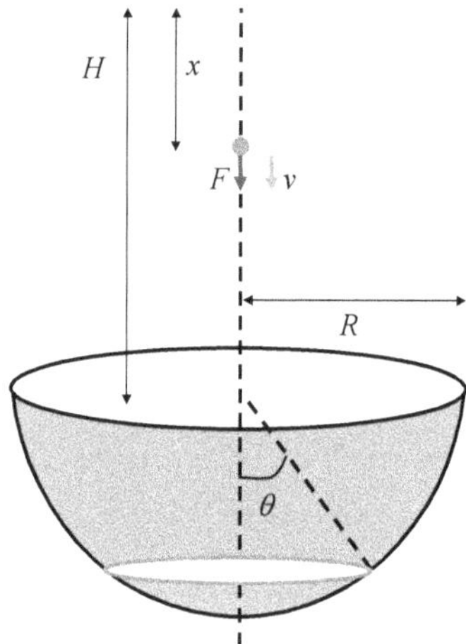

Applying Newton's second law of motion on the particle at a general time t

$$F = m(acc) \tag{1}$$

Where F is the electrostatic force acting on the particle at a general location, when it has covered a displacement x. F needs to be found by integrating the differential force of the shown ring, with appropriate limits

$$dF = \frac{q\left(\dfrac{Q}{2\pi R^2} 2\pi R \sin\theta \cdot R\,d\theta\right)}{4\pi\varepsilon_0} \frac{(H - x + R\cos\theta)}{\left[(H - x + R\cos\theta)^2 + (R\sin\theta)^2\right]^{\frac{3}{2}}} \tag{2}$$

Simplifying

$$dF = \frac{qQ\sin\theta(H - x + R\cos\theta)}{4\pi\varepsilon_0\left[(H - x)^2 + R^2 + 2(H - x)R\cos\theta\right]^{\frac{3}{2}}}\,d\theta \tag{3}$$

The net force F is therefore the definite integral of above expression with appropriate limits

$$F = \int dF$$

$$\Rightarrow F = \frac{qQ}{4\pi\varepsilon_0} \int_0^{\frac{\pi}{2}} \frac{\sin\theta\left(H - x + R\cos\theta\right)}{\left[\left(H-x\right)^2 + R^2 + 2\left(H-x\right)R\cos\theta\right]^{\frac{3}{2}}}\, d\theta \qquad (4)$$

Let $\cos\theta = U \Rightarrow -\sin\theta \cdot d\theta = dU$

$$F = -\frac{qQ}{4\pi\varepsilon_0} \int_1^0 \frac{\left(H - x + RU\right)}{\left[\left(H-x\right)^2 + R^2 + 2\left(H-x\right)RU\right]^{\frac{3}{2}}}\, dU \qquad (5)$$

Let $\left(H-x\right)^2 + R^2 + 2\left(H-x\right)RU = V^2 \Rightarrow U = \dfrac{V^2 - \left(H-x\right)^2 - R^2}{2\left(H-x\right)R}$, and also $dU = \dfrac{V \cdot dV}{\left(H-x\right)R}$

$$F = -\frac{qQ}{4\pi\varepsilon_0} \int_{H-x+R}^{\sqrt{\left(H-x\right)^2+R^2}} \frac{\left[H - x + \dfrac{V^2 - \left(H-x\right)^2 - R^2}{2\left(H-x\right)}\right]}{V^3} \frac{V \cdot dV}{\left(H-x\right)R} \qquad (6)$$

$$F = -\frac{qQ}{8\pi\varepsilon_0 \left(H-x\right)^2 R} \int_{H-x+R}^{\sqrt{\left(H-x\right)^2+R^2}} \frac{\left(H-x\right)^2 - R^2 + V^2}{V^2} \cdot dV$$

$$\Rightarrow F = -\frac{qQ}{8\pi\varepsilon_0 \left(H-x\right)^2 R} \int_{H-x+R}^{\sqrt{\left(H-x\right)^2+R^2}} \left[\frac{\left(H-x\right)^2 - R^2}{V^2} + 1\right] \cdot dV \qquad (7)$$

$$F = -\frac{qQ}{8\pi\varepsilon_0 \left(H-x\right)^2 R} \left[V - \frac{\left\{\left(H-x\right)^2 - R^2\right\}}{V}\right]_{H-x+R}^{\sqrt{\left(H-x\right)^2+R^2}} \qquad (8)$$

$$F = \frac{qQ}{8\pi\varepsilon_0 (H-x)^2 R}\left[-\frac{2R^2}{\sqrt{(H-x)^2 + R^2}} + 2R \right]$$

$$\Rightarrow F = \frac{qQ}{4\pi\varepsilon_0 (H-x)^2}\left[1 - \frac{R}{\sqrt{(H-x)^2 + R^2}} \right] \tag{9}$$

Feeding equation 9 in 1

$$\frac{qQ}{4\pi\varepsilon_0 (H-x)^2}\left[1 - \frac{R}{\sqrt{(H-x)^2 + R^2}} \right] = m\frac{vdv}{dx} \tag{10}$$

Separating the variables and integrating both sides

$$\int_0^{v_{enter}} vdv = \frac{qQ}{4\pi\varepsilon_0 m}\left[\int_0^H \frac{dx}{(H-x)^2} - \int_0^H \frac{R\cdot dx}{(H-x)^2 \sqrt{(H-x)^2 + R^2}} \right]$$

$$\Rightarrow \frac{v_{enter}^2}{2} = \frac{qQ}{4\pi\varepsilon_0 m}\left[\left(\frac{1}{H-x}\right)_{x=H} - \frac{1}{H} - \int_0^H \frac{R\cdot dx}{(H-x)^2 \sqrt{(H-x)^2 + R^2}} \right] \tag{11}$$

To evaluate the integral appearing in equation 11, let $H - x = R\tan\phi$

$$\frac{v_{enter}^2}{2} = \frac{qQ}{4\pi\varepsilon_0 m}\left[\left(\frac{1}{H-x}\right)_{x=H} - \frac{1}{H} + \int_{\tan^{-1}\frac{H}{R}}^{0} \frac{R\cdot R\sec^2\phi\cdot d\phi}{R^2\tan^2\phi\cdot R\sec\phi} \right]$$

$$\Rightarrow v_{enter}^2 = \frac{qQ}{2\pi\varepsilon_0 m}\left[\left(\frac{1}{H-x}\right)_{x=H} - \frac{1}{H} + \frac{1}{R}\int_{\tan^{-1}\frac{H}{R}}^{0} \frac{\cos\phi\cdot d\phi}{\sin^2\phi} \right] \tag{12}$$

To evaluate the integral appearing in equation 12, let $\sin\phi = W$

$$v_{enter}^2 = \frac{qQ}{2\pi\varepsilon_0 m}\left[\left(\frac{1}{H-x}\right)_{x=H} - \frac{1}{H} + \frac{1}{R}\int_{\frac{H}{\sqrt{H^2+R^2}}}^{0}\frac{dW}{W^2}\right] \tag{13}$$

$$\Rightarrow v_{enter}^2 = \frac{qQ}{2\pi\varepsilon_0 m}\left[\left(\frac{1}{H-x}\right)_{x=H} - \frac{1}{H} + \frac{\sqrt{H^2+R^2}}{RH} - \frac{1}{R}\left(\frac{1}{W}\right)_{W=0}\right]$$

Equation 13 can be expressed as

$$v_{enter}^2 = \frac{qQ}{2\pi\varepsilon_0 m}\left[\left(\frac{1}{H-x}\right)_{x=H} - \frac{1}{H} + \frac{\sqrt{H^2+R^2}}{RH} - \frac{1}{R}\left(\frac{\sqrt{(H-x)^2+R^2}}{H-x}\right)_{x=H}\right] \tag{14}$$

$$\Rightarrow v_{enter}^2 = \frac{qQ}{2\pi\varepsilon_0 m}\left[\frac{1}{R}\left(\frac{R-\sqrt{(H-x)^2+R^2}}{H-x}\right)_{x=H} - \frac{1}{H} + \frac{\sqrt{H^2+R^2}}{RH}\right]$$

The term $\left(\dfrac{R-\sqrt{(H-x)^2+R^2}}{R(H-x)}\right)_{x=H}$ is indeterminate ($\frac{0}{0}$ form). So its limit needs to be taken

$$\lim\left(\frac{R-\sqrt{(H-x)^2+R^2}}{R(H-x)}\right)_{x=H} = \lim\left(\frac{-\dfrac{(H-x)}{\sqrt{(H-x)^2+R^2}}}{R}\right)_{x=H} = 0 \tag{15}$$

Feeding equation 15 in 14 gets

$$v_{enter} = \sqrt{\frac{qQ\left(\sqrt{H^2+R^2}-R\right)}{2\pi\varepsilon_0 mRH}} \tag{16}$$

Question 12

2 uniformly charged linear bodies are kept parallel to each other at a separation d. If the total charge, mass and length of each are Q,m,L, find the force exerted by one on the other. Next, suppose one of them is fixed while other is set free at $t=0$. Find the velocity of the movable one as a function of its displacement. Consider only electrostatic forces. The permittivity of free space is ε_0

SOLUTION

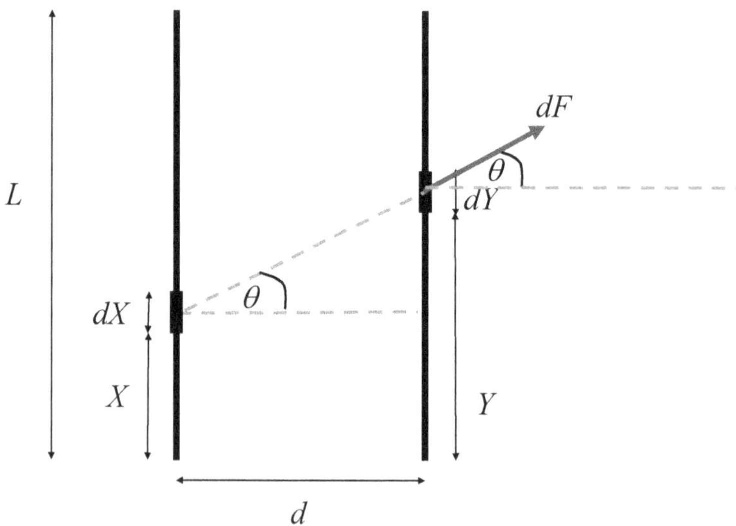

The magnitude of the differential force exerted by a general, differential point charge lying on one line on a general, differential point charge lying on the other line is

$$dF = \frac{dq_1 \cdot dq_2}{4\pi\varepsilon_0 \left[(Y-X)^2 + d^2 \right]} \tag{1}$$

Where dq_1 and dq_2 are the magnitudes of these point charges. Since , $dq_1 = \frac{Q}{L}dX$, $dq_2 = \frac{Q}{L}dY$

$$dF = \frac{Q^2 dX \cdot dY}{4\pi\varepsilon_0 L^2 \left[(Y-X)^2 + d^2 \right]} \tag{2}$$

This force has two components, one normal to the second line and other along it. They are respectively,

$$dF_x = dF\cos\theta$$

$$\Rightarrow dF_x = \frac{Q^2 \cos\theta dX \cdot dY}{4\pi\varepsilon_0 L^2 \left[(Y-X)^2 + d^2 \right]}$$

$$\Rightarrow dF_x = \frac{Q^2 \cdot d \cdot dX \cdot dY}{4\pi\varepsilon_0 L^2 \left[(Y-X)^2 + d^2 \right]^{1.5}} \tag{3}$$

$$dF_y = dF\sin\theta$$

$$\Rightarrow dF_y = \frac{Q^2 \sin\theta \cdot dX \cdot dY}{4\pi\varepsilon_0 L^2 \left[(Y-X)^2 + d^2 \right]}$$

$$\Rightarrow dF_y = \frac{Q^2 (Y-X) dX \cdot dY}{4\pi\varepsilon_0 L^2 \left[(Y-X)^2 + d^2 \right]^{1.5}} \tag{4}$$

The horizontal and vertical components of the net force between the lines are

$$F_x = \int dF_x$$

$$\Rightarrow F_x = \int_0^L \int_0^L \frac{Q^2 \cdot d \cdot dX \cdot dY}{4\pi\varepsilon_0 L^2 \left[(Y-X)^2 + d^2 \right]^{1.5}} \tag{5}$$

$$F_y = \int dF_y$$

$$\Rightarrow F_y = \int_0^L \int_0^L \frac{Q^2 (Y-X) dX \cdot dY}{4\pi\varepsilon_0 L^2 \left[(Y-X)^2 + d^2 \right]^{1.5}} \tag{6}$$

The double integrals appearing in equations 5 and 6 are evaluated as follows: First integrate with respect to Y, treating X as a constant and then with respect to X. In equation 5, for the integral with respect to Y, let $Y - X = d \tan\phi \Rightarrow dY = d \sec^2 \phi \cdot d\phi$

$$F_x = \frac{Q^2 d}{4\pi\varepsilon_0 L^2} \int_0^L \left[\int_{\tan^{-1} \frac{-X}{d}}^{\tan^{-1} \frac{L-X}{d}} \frac{d \sec^2 \phi \cdot d\phi}{d^3 \sec^3 \phi} \right] dX \tag{7}$$

Simplifying

$$F_x = \frac{Q^2}{4\pi d\varepsilon_0 L^2} \int_0^L \left[\frac{L-X}{\sqrt{d^2 + (L-X)^2}} + \frac{X}{\sqrt{d^2 + X^2}} \right] dX \tag{8}$$

$$F_x = -\frac{Q^2}{4\pi d\varepsilon_0 L^2} \int_{\sqrt{d^2+L^2}}^{d} dU + \frac{Q^2}{4\pi\varepsilon_0 dL^2} \int_d^{\sqrt{d^2+L^2}} dV \tag{9}$$

$$\Rightarrow F_x = \frac{Q^2}{2\pi d\varepsilon_0 L^2} \left[\sqrt{d^2 + L^2} - d \right]$$

In equation 6, for the integral with respect to y, let $(Y-X)^2 + d^2 = W^2 \Rightarrow (Y-X) dY = W \cdot dW$

$$F_y = \frac{Q^2}{4\pi\varepsilon_0 L^2} \int_0^L \left[\int_{\sqrt{X^2+d^2}}^{\sqrt{(L-X)^2+d^2}} \frac{dW}{W^2} \right] dX \tag{10}$$

Simplifying

$$F_y = \frac{Q^2}{4\pi\varepsilon_0 L^2} \int_0^L \left[\frac{1}{\sqrt{X^2+d^2}} - \frac{1}{\sqrt{(L-X)^2+d^2}} \right] dX \tag{11}$$

$$F_y = \frac{Q^2}{4\pi\varepsilon_0 L^2} \left[\ln\left|X + \sqrt{X^2+d^2}\right| + \ln\left|L - X + \sqrt{(L-X)^2+d^2}\right| \right]_0^L \tag{12}$$

$$\Rightarrow F_y = 0$$

Therefore, the force by one line on the other is completely along the direction normal to them. So, when one of the lines is set free, it shall move in this direction, away from the other line (repulsion). So we have one dimensional motion. When it has covered a displacement x, the expression of force on it will be the one given in equation 9, with d replaced by $d+x$. Applying Newton's second law of motion on the moving line, at a general time t

$$\frac{Q^2}{2\pi\varepsilon_0 (d+x) L^2} \left[\sqrt{(d+x)^2 + L^2} - (d+x) \right] = mv\frac{dv}{dx} \tag{13}$$

Separating the variables and integrating both sides

$$\int_0^v v \cdot dv = \int_0^x \frac{Q^2 \left[\sqrt{(d+x)^2 + L^2} - (d+x) \right]}{2\pi\varepsilon_0 (d+x) mL^2} dx \tag{14}$$

$$\frac{v^2}{2} = \frac{Q^2}{2\pi\varepsilon_0 mL^2} \int_0^x \left[\frac{\sqrt{L^2 + (d+x)^2}}{d+x} - 1 \right] dx \tag{15}$$

Let $d + x = L\tan\phi \Rightarrow dx = L\sec^2\phi \cdot d\phi$

$$\frac{v^2}{2} = \frac{Q^2}{2\pi\varepsilon_0 mL^2} \left[\int_{\tan^{-1}\frac{d}{L}}^{\tan^{-1}\frac{d+x}{L}} \frac{L\sec\phi}{L\tan\phi} L\sec^2\phi \cdot d\phi - x \right] \tag{16}$$

$$\frac{v^2}{2} = \frac{Q^2}{2\pi\varepsilon_0 mL^2} \left[\int_{\tan^{-1}\frac{d}{L}}^{\tan^{-1}\frac{d+x}{L}} \frac{L \cdot d\phi}{\sin\phi\cos^2\phi} - x \right] \tag{17}$$

$$\Rightarrow \frac{v^2}{2} = \frac{Q^2}{2\pi\varepsilon_0 mL^2} \left[\int_{\tan^{-1}\frac{d}{L}}^{\tan^{-1}\frac{d+x}{L}} \frac{L\sin\phi \cdot d\phi}{\sin^2\phi\cos^2\phi} - x \right]$$

Let $\cos\phi = P \Rightarrow -\sin\phi \cdot d\phi = dP$

$$\frac{v^2}{2} = \frac{Q^2}{2\pi\varepsilon_0 mL^2} \left[-L \int_{\cos\left(\tan^{-1}\frac{d}{L}\right)}^{\cos\left(\tan^{-1}\frac{d+x}{L}\right)} \frac{dP}{(1-P^2)P^2} - x \right] \tag{18}$$

$$\frac{v^2}{2} = \frac{Q^2}{2\pi\varepsilon_0 m L^2}\left[-L\int_{\substack{\cos\left(\tan^{-1}\frac{d}{L}\right)\\-x}}^{\substack{\cos\left(\tan^{-1}\frac{d+x}{L}\right)}}\left(\frac{1}{P^2}+\frac{1}{1-P^2}\right)dP\right] \tag{19}$$

$$\frac{v^2}{2} = \frac{Q^2}{2\pi\varepsilon_0 m L}\left[\frac{1}{2}\ln\left|\frac{\sqrt{L^2+\left(d+x\right)^2}+L}{\sqrt{L^2+\left(d+x\right)^2}-L}\cdot\frac{\sqrt{L^2+d^2}-L}{\sqrt{L^2+d^2}+L}\right|+\frac{\sqrt{L^2+\left(d+x\right)^2}}{L}-\frac{\sqrt{L^2+d^2}}{L}\right] \tag{20}$$

$$v = \sqrt{\frac{Q^2}{\pi\varepsilon_0 m L}\left[\frac{1}{2}\ln\left|\frac{\sqrt{L^2+\left(d+x\right)^2}+L}{\sqrt{L^2+\left(d+x\right)^2}-L}\cdot\frac{\sqrt{L^2+d^2}-L}{\sqrt{L^2+d^2}+L}\right|+\frac{\sqrt{L^2+\left(d+x\right)^2}}{L}-\frac{\sqrt{L^2+d^2}}{L}\right]} \tag{21}$$

Question 13

❖ ❖ ❖

2 parallel rails are connected at one end, with a rod placed over them as shown. A spatially variable magnetic field $B = B_0 e^{-kX}$ exists into the plane of paper in the region. The rod is imparted a rightward velocity v_0 at time $t=0$. Find the rod's displacement as a function of time. The mass of rod is m, its electrical resistance is R, and resistance of rails and the connecting bar (for the rails) is ignorable. Further, air resistance is neglected. The entire assembly other than the rod is kept fixed.

SOLUTION

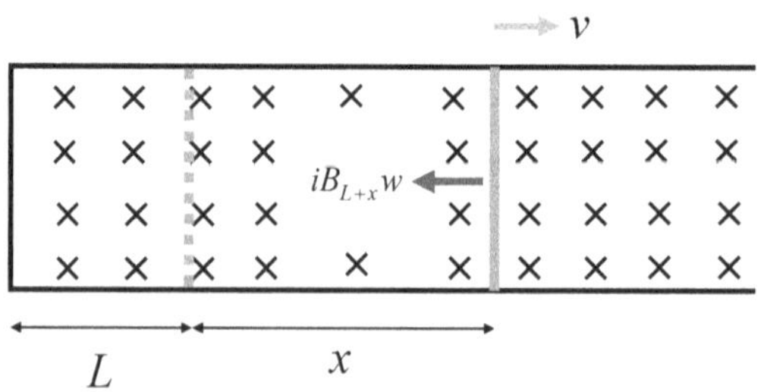

As the rod begins moving, the region between rod and the connected end of the rails becomes a closed loop with varying magnetic flux. An electromotive force is therefore developed in this closed loop. Since the loop is metallic, current will flow in it. A current carrying wire experiences a magnetic force when placed in a magnetic field perpendicular to its plane. Thus, each of the 4 constituents of the loop will experience a magnetic force. Since 3 of them are kept fixed, only the 4th (the rod) will get affected by this force.

Applying Newton's second law of motion on the rod at a general time t, when it has gotten displaced by an amount x

$$\vec{F}_{magnetic} = m\left(\vec{acc}\right) \tag{1}$$

The magnetic force on the rod is given by

$$F_{magnetic} = \int i \cdot \vec{dl} \times \vec{B}_{L+x} = \int i \cdot \left(dl\,\hat{j}\right) \times \left(-B_{L+x}\,\hat{k}\right) = -\int \left(i \cdot dl \cdot B_{L+x}\right)\hat{i} = -iB_{L+x}\int dl\,\hat{i} = -iB_{L+x}w\,\hat{i} \tag{2}$$

Feeding equation 2 in 1

$$-iB_{L+x}w = m\left(acc\right)$$
$$\Rightarrow acc = -\frac{iB_0 e^{-k(L+x)}w}{m} \tag{3}$$

Instantaneous current in the loop will be given by Ohm's law

$$i = \frac{\varepsilon}{R} \tag{4}$$

Instantaneous EMF in the loop will be given by Faraday's law of electromagnetic induction

$$|\varepsilon| = \left|\frac{d\phi}{dt}\right| \tag{5}$$

The instantaneous magnetic flux through the loop is required to calculate the above, which is given as

$$\phi = \int\limits_0^{L+x} B_X \left(w \cdot dX \right) = \int\limits_0^{L+x} B_0 e^{-kX} \left(w \cdot dX \right)$$

$$\Rightarrow \phi = \frac{B_0 w \left\{ 1 - e^{-k(x+L)} \right\}}{k} \tag{6}$$

Feeding equation 6 in 5 and the resulting equation in equation 4

$$i = \frac{B_0 w e^{-k(x+L)}}{R} \frac{dx}{dt} \tag{7}$$

Feeding equation 7 in equation 3

$$acc = -\frac{B_0^2 w^2 e^{-2k(x+L)}}{mR} v \tag{8}$$

Equation 8 can be expressed as

$$\frac{v\,dv}{dx} = -\frac{B_0^2 w^2 e^{-2k(x+L)}}{mR} v$$

$$\Rightarrow \frac{dv}{dx} = -\frac{B_0^2 w^2 e^{-2k(x+L)}}{mR} \tag{9}$$

Separating the variables and integrating both sides

$$\int\limits_{v_0}^{v} dv = \int\limits_0^{x} -\frac{B_0^2 w^2 e^{-2k(x+L)}}{mR} dx$$

$$\Rightarrow v - v_0 = -\frac{B_0^2 w^2 e^{-2kL} \left(1 - e^{-2kx} \right)}{2kmR} \tag{10}$$

Equation 10 can be expressed as

$$\frac{dx}{dt} = \alpha + \beta e^{-2kx} \tag{11}$$

Where $\alpha = v_0 - \dfrac{B_0^2 w^2 e^{-2kL}}{2kmR}$, $\beta = \dfrac{B_0^2 w^2 e^{-2kL}}{2kmR}$

Separating the variables and integrating both sides

$$\int_0^x \frac{dx}{\alpha + \beta e^{-2kx}} = \int_0^t dt \tag{12}$$

Let $e^{-2kx} = U \Rightarrow -2ke^{-2kx}dx = dU$

$$-\int_1^{e^{-2kx}} \frac{dU}{2kU\left(\alpha + \beta U\right)} = t$$

$$\Rightarrow \int_1^{e^{-2kx}} \frac{dU}{U\left(\alpha + \beta U\right)} = -2kt \tag{13}$$

Simplifying

$$\int_1^{e^{-2kx}} \frac{\alpha \cdot dU}{\beta U\left(\alpha + \beta U\right)} = -\frac{2k\alpha t}{\beta}$$

$$\Rightarrow \int_1^{e^{-2kx}} \left(\frac{1}{\beta U} - \frac{1}{\alpha + \beta U}\right) dU = -\frac{2k\alpha t}{\beta} \tag{14}$$

$$\left[\frac{1}{\beta}\ln|U| - \frac{1}{\beta}\ln|\alpha + \beta U|\right]_1^{e^{-2kx}} = -\frac{2k\alpha t}{\beta} \tag{15}$$

$$\ln\left|\frac{\alpha + \beta}{\alpha e^{2kx} + \beta}\right| = -2k\alpha t$$

$$\Rightarrow \frac{\alpha + \beta}{\alpha e^{2kx} + \beta} = e^{-2k\alpha t} \tag{16}$$

$$\Rightarrow x = \frac{1}{2k}\ln\left|\frac{\left(\alpha + \beta\right)e^{2k\alpha t} - \beta}{\alpha}\right|$$

Expanding back $\alpha,\ \beta$

$$x = \frac{1}{2k}\ln\left|\frac{v_0 e^{\left(2kv_0 - \frac{B_0^2 w^2 e^{-2kL}}{mR}\right)t} - \frac{B_0^2 w^2 e^{-2kL}}{2kmR}}{v_0 - \frac{B_0^2 w^2 e^{-2kL}}{2kmR}}\right| \tag{17}$$

Question 14

❖ ❖ ❖

Find velocity of the rod as a function of its displacement in the previous question if the magnetic field in the region varies as $B = B_0 XY$. Further, consider air drag to be present as per the quadratic law: $\vec{F}_{air\ drag} = -k|\vec{v}|^2\,\hat{v}$.

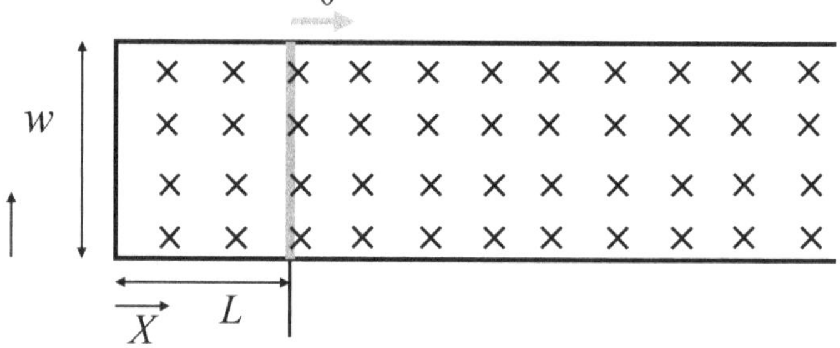

SOLUTION

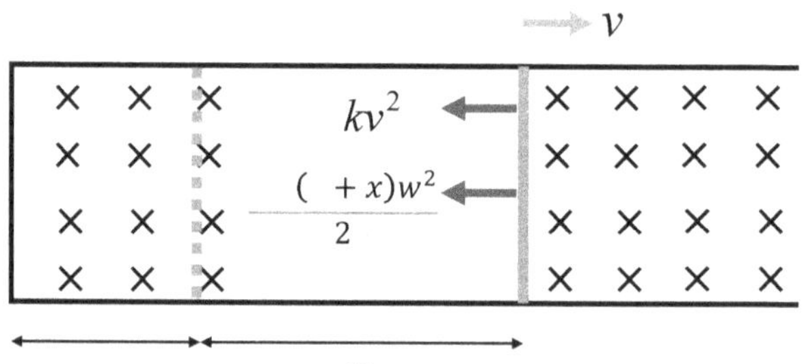

Applying Newton's second law of motion on the rod at a general time t, when it has gotten displaced by an amount x

$$\vec{F}_{magnetic} + \vec{F}_{air\ drag} = m(\vec{acc}) \tag{1}$$

The magnetic force on the rod is given by

$$F_{magnetic} = \int_0^w i \cdot \vec{dY} \times \vec{B}_{X,Y} = \int_0^w i \cdot \left(dY\ \hat{j} \right) \times \left(-B_{X,Y}\ \hat{k} \right) = -\int_0^w \left(iB_{X,Y} \cdot dY \right) \hat{i} = -\int_0^w \left(iB_0(L+x)Y \cdot dY \right) \hat{i}$$

$$\Rightarrow F_{magnetic} = -iB_0(L+x)\int_0^w (Y \cdot dY)\ \hat{i} = -\frac{iB_0(L+x)w^2}{2}\ \hat{i} \tag{2}$$

Feeding equation 2 and given expression of air drag in equation 1

$$acc = -\frac{iB_0(L+x)w^2}{2m} - \frac{k}{m}v^2 \tag{3}$$

Instantaneous current in the loop will be given by Ohm's law

$$i = \frac{\varepsilon}{R} \tag{4}$$

Instantaneous EMF in the loop will be given by Faraday's law of electromagnetic induction

$$|\varepsilon| = \left| \frac{d\phi}{dt} \right| \tag{5}$$

The instantaneous magnetic flux through the loop is required to calculate the above, which is given as

$$\phi = \int_0^{L+x} \left(\int_0^w B_{X,Y} dY \right) dX = \int_0^{L+x} \left(\int_0^w B_0 XY \cdot dY \right) dX = \int_0^{L+x} B_0 X \left(\int_0^w YdY \right) dX = \int_0^{L+x} \frac{B_0 Xw^2}{2} dX = \frac{B_0(L+x)^2 w^2}{4} \tag{6}$$

Feeding equation 6 in 5 and the resulting equation in equation 4

$$i = \frac{B_0(L+x)w^2 v}{2R} \tag{7}$$

Feeding equation 7 in equation 3

$$v\frac{dv}{dx} = -\frac{B_0^2 (L+x)^2 w^4 v}{4mR} - \frac{k}{m}v^2$$

$$\Rightarrow \frac{dv}{dx} + \frac{k}{m}v = -\frac{B_0^2 (L+x)^2 w^4}{4mR} \tag{8}$$

Equation 8 is a first order linear ordinary differential equation whose solution can be given by the method of integrating factor as

$$v(\text{I}\cdot\text{F}) = \int -\frac{B_0^2 (L+x)^2 w^4}{4mR}(\text{I}\cdot\text{F})dx + C_0 \tag{9}$$

Where $\text{I}\cdot\text{F} = e^{\int \frac{k}{m}dx} = e^{\frac{kx}{m}}$ and C_0 is an arbitrary integration constant

$$ve^{\frac{kx}{m}} = -\left[\frac{B_0^2 w^4}{4mR}\int (L+x)^2 e^{\frac{kx}{m}}dx\right] + C_0 \tag{10}$$

The integral appearing on the right-hand side of equation 10 can be evaluated using integration by parts as shown below

$$\int (L+x)^2 e^{\frac{kx}{m}}dx = (L+x)^2 \int e^{\frac{kx}{m}}dx - \int \left(2(L+x)\int e^{\frac{kx}{m}}dx\right)dx$$

$$\Rightarrow \int (L+x)^2 e^{\frac{kx}{m}}dx = \frac{m(L+x)^2}{k}e^{\frac{kx}{m}} - \frac{2m}{k}\int (L+x)e^{\frac{kx}{m}}dx$$

$$\Rightarrow \int (L+x)^2 e^{\frac{kx}{m}}dx = \frac{m(L+x)^2}{k}e^{\frac{kx}{m}} - \frac{2m}{k}\left[\frac{m(L+x)}{k}e^{\frac{kx}{m}} - \int \frac{m}{k}e^{\frac{kx}{m}}dx\right] \tag{11}$$

$$\Rightarrow \int (L+x)^2 e^{\frac{kx}{m}}dx = \frac{m(L+x)^2}{k}e^{\frac{kx}{m}} - \frac{2m}{k}\left[\frac{m(L+x)}{k}e^{\frac{kx}{m}} - \frac{m^2}{k^2}e^{\frac{kx}{m}}\right]$$

$$\Rightarrow \int (L+x)^2 e^{\frac{kx}{m}}dx = \frac{me^{\frac{kx}{m}}}{k}\left[(L+x)^2 - \frac{2m(L+x)}{k} + \frac{2m^2}{k^2}\right]$$

Feeding equation 11 in 10

$$ve^{\frac{kx}{m}} = -\left[\frac{B_0^2 w^4}{4mR}\frac{me^{\frac{kx}{m}}}{k}\left\{(L+x)^2 - \frac{2m(L+x)}{k} + \frac{2m^2}{k^2}\right\}\right] + C_0$$

$$\Rightarrow v = C_0 e^{-\frac{kx}{m}} - \frac{B_0^2 w^4}{4kR}\left\{(L+x)^2 - \frac{2m(L+x)}{k} + \frac{2m^2}{k^2}\right\}$$

(12)

The constant C_0 needs to be found using the initial condition: $v_{x=0} = v_0$

$$C_0 = v_0 + \frac{B_0^2 w^4\left(L^2 - \frac{2mL}{k} + \frac{2m^2}{k^2}\right)}{4kR}$$

(13)

Feeding equation 13 in 12

$$v = \left[\left\{v_0 + \frac{B_0^2 w^4\left(L^2 - \frac{2mL}{k} + \frac{2m^2}{k^2}\right)}{4kR}\right\}e^{-\frac{kx}{m}} - \left[\frac{B_0^2 w^4}{4kR}\left\{(L+x)^2 - \frac{2m(L+x)}{k} + \frac{2m^2}{k^2}\right\}\right]\right]$$

(14)

Question 15

❖ ❖ ❖

A charged disc of radius R rotates on a rough floor (friction coefficient for the disc floor pair is μ). Charge and mass are distributed on the disc uniformly [charge per unit area= σ_1 and mass per unit area = σ_2]. A point charge q is located above the disc at a height h from it, along its axis and is kept fixed. The permittivity of free space is ε_0 and acceleration due to gravity on the Earth's surface is g. Find the ratio $\frac{\sigma_2}{\sigma_1}$ for the disc to keep rotating indefinitely. Neglect air resistance.

SOLUTION

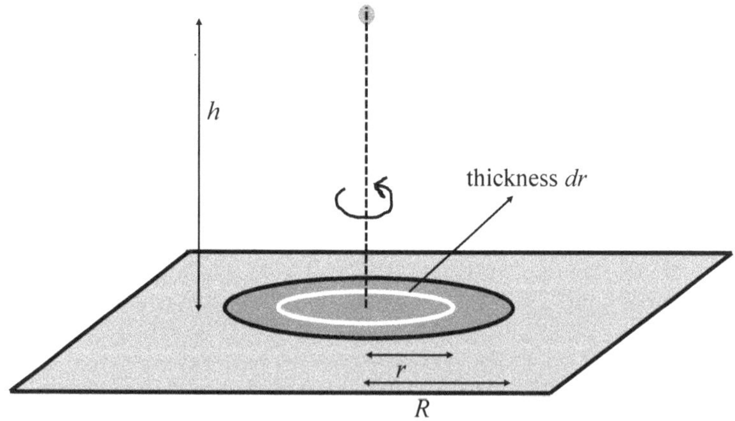

For a differential ring of radius r and thickness dr, the differential force due to point charge q (acting upwards) is

$$dF = \frac{q\left(\sigma_1.2\pi r \cdot dr\right)h}{4\pi\varepsilon_0 \left(r^2 + h^2\right)^{\frac{3}{2}}} \tag{1}$$

In the vertical direction, for equilibrium of the differential ring

$$dF + dN = dm.g \tag{2}$$

Where dN is the normal reaction on this differential ring and dm is it mass.

Feeding equation 1 in 2 and using the given expression of areal mass density

$$\frac{qh\sigma_1\pi r \cdot dr}{2\pi\varepsilon_0 \left(r^2 + h^2\right)^{\frac{3}{2}}} + dN = 2\pi\sigma_2 gr \cdot dr$$

$$\Rightarrow dN = 2\pi\sigma_2 gr \cdot dr - \frac{qh\sigma_1 r \cdot dr}{2\varepsilon_0 \left(r^2 + h^2\right)^{\frac{3}{2}}} \tag{3}$$

The torque of the frictional force acting on the differential ring is

$$d\tau = \mu \cdot dN \cdot r = 2\pi\sigma_2 g\mu r^2 \cdot dr - \frac{qh\sigma_1 \mu r^2 \cdot dr}{2\varepsilon_0 \left(r^2 + h^2\right)^{\frac{3}{2}}} \tag{4}$$

The net torque on the rotating disc is

$$\tau = \int d\tau = 2\pi\sigma_2 g\mu \int_0^R r^2 \cdot dr - \frac{qh\sigma_1 \mu}{2\varepsilon_0} \int_0^R \frac{r^2 \cdot dr}{\left(r^2 + h^2\right)^{\frac{3}{2}}} \tag{5}$$

To evaluate the right-hand side integral appearing in equation 5, let $r = h\tan\theta \Rightarrow dr = h\sec^2\theta \cdot d\theta$.

$$\tau = \frac{2\pi\sigma_2\mu gR^3}{3} - \frac{\mu q\sigma_1 h}{2\varepsilon_0}\int_0^{\tan^{-1}\frac{R}{h}}\frac{h^3\tan^2\theta\sec^2\theta\cdot d\theta}{h^3\sec^3\theta}$$

$$\Rightarrow \tau = \frac{2\pi\sigma_2\mu gR^3}{3} - \frac{\mu q\sigma_1 h}{2\varepsilon_0}\int_0^{\tan^{-1}\frac{R}{h}}\tan^2\theta\cos\theta\cdot d\theta \tag{6}$$

$$\Rightarrow \tau = \frac{2\pi\sigma_2\mu gR^3}{3} - \frac{\mu q\sigma_1 h}{2\varepsilon_0}\int_0^{\tan^{-1}\frac{R}{h}}\frac{\sin^2\theta\cos\theta\cdot d\theta}{1-\sin^2\theta}$$

Let $\sin\theta = V \Rightarrow \cos\theta\cdot d\theta = dV$

$$\tau = \frac{2\pi\sigma_2\mu gR^3}{3} - \frac{\mu q\sigma_1 h}{2\varepsilon_0}\int_0^{\sin\left(\tan^{-1}\frac{R}{h}\right)}\frac{V^2\cdot dV}{1-V^2}$$

$$\Rightarrow \tau = \frac{2\pi\sigma_2\mu gR^3}{3} + \frac{\mu q\sigma_1 h}{2\varepsilon_0}\int_0^{\sin\left(\tan^{-1}\frac{R}{h}\right)}\frac{\left(1-V^2-1\right)\cdot dV}{1-V^2} \tag{7}$$

$$\Rightarrow \tau = \frac{2\pi\sigma_2\mu gR^3}{3} + \frac{\mu q\sigma_1 h}{2\varepsilon_0}\int_0^{\sin\left(\tan^{-1}\frac{R}{h}\right)}dV - \frac{\mu q\sigma_1 h}{2\varepsilon_0}\int_0^{\sin\left(\tan^{-1}\frac{R}{h}\right)}\frac{dV}{1-V^2}$$

Using the standard result: $\int\frac{dZ}{S^2-Z^2} = \frac{1}{2S}\ln\left|\frac{Z+S}{Z-S}\right|$

$$\tau = \frac{2\pi\sigma_2\mu gR^3}{3} + \frac{\mu q\sigma_1 hR}{2\varepsilon_0\sqrt{R^2+h^2}} + \frac{\mu q\sigma_1 h}{4\varepsilon_0}\ln\left|\frac{R-\sqrt{R^2+h^2}}{R+\sqrt{R^2+h^2}}\right| \tag{8}$$

For the disc to keep rotating forever, net torque must be zero

$$\frac{2\pi\sigma_2\mu gR^3}{3} + \frac{\mu q\sigma_1 hR}{2\varepsilon_0\sqrt{R^2+h^2}} + \frac{\mu q\sigma_1 h}{4\varepsilon_0}\ln\left|\frac{R-\sqrt{R^2+h^2}}{R+\sqrt{R^2+h^2}}\right| = 0$$

$$\Rightarrow \frac{2\pi\sigma_2\mu gR^3}{3} = \frac{\mu q\sigma_1 h}{4\varepsilon_0}\ln\left|\frac{R+\sqrt{R^2+h^2}}{R-\sqrt{R^2+h^2}}\right| - \frac{\mu q\sigma_1 hR}{2\varepsilon_0\sqrt{R^2+h^2}} \tag{9}$$

Using equation 9, the required ratio is found out to be

$$\frac{\sigma_2}{\sigma_1} = \frac{\mu q h}{2\varepsilon_0}\left[\frac{\frac{1}{2}\ln\left|\frac{R+\sqrt{R^2+h^2}}{R-\sqrt{R^2+h^2}}\right| - \frac{R}{\sqrt{R^2+h^2}}}{\frac{2\pi\mu g R^3}{3}}\right] \tag{10}$$

Question 16

A particle of mass m, and charge Q rests at the origin. Two identical rectangular hyperbolic curves containing uniformly distributed negative charge exist in the XY plane as shown. Find the value of uniform, charge per unit length on these curves for which the charge Q remains in equilibrium. The permittivity of free space is ε_0. The acceleration due to gravity on Earth's surface is g. Bouyancy is neglected

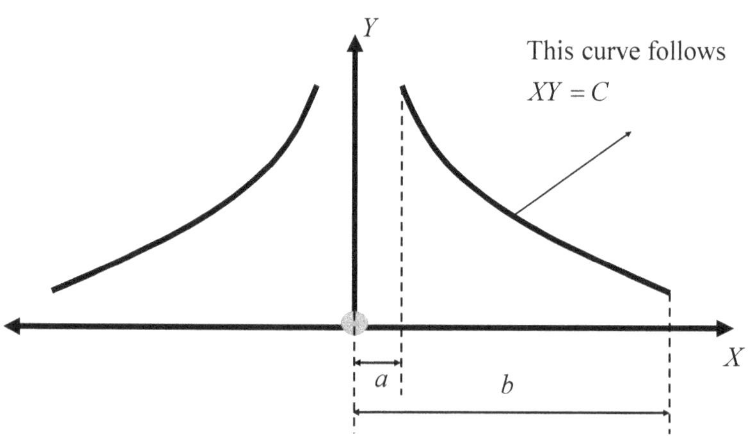

SOLUTION

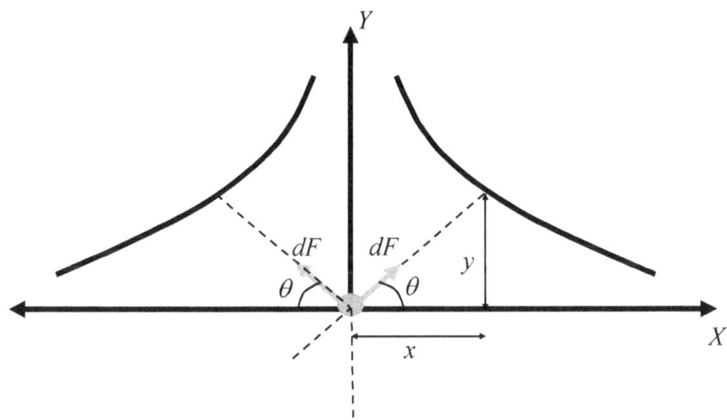

The magnitude of the differential force acting on the charge Q because of a differential element on the right curve whose coordinates are (x,y) is

$$|dF| = \frac{1}{4\pi\varepsilon_0} \frac{Q(\lambda \cdot dL)}{x^2 + y^2} \tag{1}$$

Where dL is a differential length along the curve. By symmetry, a similar differential force vector is caused by the element on the other curve whose coordinates are $(-x,y)$. The horizontal components of these two vectors shall cancel out and the vertical ones shall add up. Therefore, the net vertical force acting on the charge Q will be

$$F_{net} = \int 2|dF|\sin\theta = \frac{Q\lambda}{2\pi\varepsilon_0} \int \frac{dL \sin\theta}{x^2 + y^2} \tag{2}$$

From geometry,

$$dL = \sqrt{(dx)^2 + (dy)^2} = dx\sqrt{1 + \left(\frac{dy}{dx}\right)^2} \tag{3}$$

$$\sin \theta = \frac{y}{\sqrt{x^2 + y^2}} \tag{4}$$

Feeding equations 3 and 4 in 2

$$F_{net} = \frac{Q\lambda}{2\pi\varepsilon_0} \int_a^b \frac{y\sqrt{1+\left(\frac{dy}{dx}\right)^2}}{\left(x^2 + y^2\right)^{\frac{3}{2}}} dx \tag{5}$$

The coordinates (x,y) shall satisfy the given curve equation. Therefore,

$$y = \frac{C}{x} \Rightarrow \frac{dy}{dx} = \frac{-C}{x^2} \tag{6}$$

Feeding equation 6 in 5

$$F_{net} = \frac{Q\lambda}{2\pi\varepsilon_0} \int_a^b \frac{\frac{C}{x}\sqrt{1+\frac{C^2}{x^4}}}{\left(x^2 + \frac{C^2}{x^2}\right)^{\frac{3}{2}}} dx \tag{7}$$

$$\Rightarrow F_{net} = \frac{Q\lambda C}{2\pi\varepsilon_0} \int_a^b \frac{dx}{x^4 + C^2}$$

Simplifying

$$F_{net} = \frac{Q\lambda C}{4\pi\varepsilon_0} \int_a^b \frac{\frac{2}{x^2} \cdot dx}{x^2 + \frac{C^2}{x^2}} \tag{8}$$

$$F_{net} = \frac{Q\lambda}{4\pi\varepsilon_0} \int_a^b \frac{\frac{2C}{x^2} \cdot dx}{x^2 + \frac{C^2}{x^2} + 2C - 2C} \tag{9}$$

$$F_{net} = \frac{Q\lambda}{4\pi\varepsilon_0} \int_a^b \frac{\left(1+\frac{C}{x^2}\right) - \left(1-\frac{C}{x^2}\right) \cdot dx}{x^2 + \frac{C^2}{x^2} + 2C - 2C} \tag{10}$$

$$F_{net} = \frac{Q\lambda}{4\pi\varepsilon_0} \int_a^b \frac{\left(1+\frac{C}{x^2}\right)dx}{\left(x-\frac{C}{x}\right)^2 + \left(\sqrt{2C}\right)^2} - \frac{Q\lambda}{4\pi\varepsilon_0} \int_a^b \frac{\left(1-\frac{C}{x^2}\right)dx}{\left(x+\frac{C}{x}\right)^2 - \left(\sqrt{2C}\right)^2} \tag{11}$$

For the leftward integral, let $x - \frac{C}{x} = U \Rightarrow \left(1+\frac{C}{x^2}\right)dx = dU$

For the rightward integral, let $x + \frac{C}{x} = V \Rightarrow \left(1-\frac{C}{x^2}\right)dx = dV$

$$F_{net} = \frac{Q\lambda}{4\pi\varepsilon_0} \int_{a-\frac{C}{a}}^{b-\frac{C}{b}} \frac{dU}{U^2 + \left(\sqrt{2C}\right)^2} - \frac{Q\lambda}{4\pi\varepsilon_0} \int_{a+\frac{C}{a}}^{b+\frac{C}{b}} \frac{dV}{V^2 - \left(\sqrt{2C}\right)^2} \tag{12}$$

$$F_{net} = \frac{Q\lambda}{4\pi\varepsilon_0\sqrt{2C}} \cdot \left[\tan^{-1}\left(\frac{b-\frac{C}{b}}{\sqrt{2C}}\right) - \tan^{-1}\left(\frac{a-\frac{C}{a}}{\sqrt{2C}}\right)\right] - \frac{Q\lambda}{4\pi\varepsilon_0} \cdot \frac{1}{2\sqrt{2C}} \ln\left|\frac{b+\frac{C}{b}-\sqrt{2C}}{b+\frac{C}{b}+\sqrt{2C}} \cdot \frac{a+\frac{C}{a}+\sqrt{2C}}{a+\frac{C}{a}-\sqrt{2C}}\right| \tag{13}$$

This net force suffered by the point charge Q must be balanced by its weight for its equilibrium

$$\frac{Q\lambda}{4\pi\varepsilon_0\sqrt{2C}} \cdot \left[\tan^{-1}\left(\frac{b-\frac{C}{b}}{\sqrt{2C}}\right) - \tan^{-1}\left(\frac{a-\frac{C}{a}}{\sqrt{2C}}\right)\right] - \frac{Q\lambda}{4\pi\varepsilon_0} \cdot \frac{1}{2\sqrt{2C}} \ln\left|\frac{b+\frac{C}{b}-\sqrt{2C}}{b+\frac{C}{b}+\sqrt{2C}} \cdot \frac{a+\frac{C}{a}+\sqrt{2C}}{a+\frac{C}{a}-\sqrt{2C}}\right| = mg$$

$$\Rightarrow \lambda = \frac{mg}{\frac{Q}{4\pi\varepsilon_0\sqrt{2C}} \cdot \left[\tan^{-1}\left(\frac{b-\frac{C}{b}}{\sqrt{2C}}\right) - \tan^{-1}\left(\frac{a-\frac{C}{a}}{\sqrt{2C}}\right)\right] - \frac{Q}{4\pi\varepsilon_0} \cdot \frac{1}{2\sqrt{2C}} \ln\left|\frac{b+\frac{C}{b}-\sqrt{2C}}{b+\frac{C}{b}+\sqrt{2C}} \cdot \frac{a+\frac{C}{a}+\sqrt{2C}}{a+\frac{C}{a}-\sqrt{2C}}\right|} \tag{14}$$

Question 17

A chain of length L, uniformly charged (total charge $-Q$) has a fraction f kept on a table and the rest overhangs. An electric field $E = \dfrac{E_0}{y^2}$ exists in the region as shown. The static friction coefficient between the table and chain is μ Find the minimum value of f for chain's equilibrium to not be disturbed. Neglect gravitational, air resistance and buoyant forces

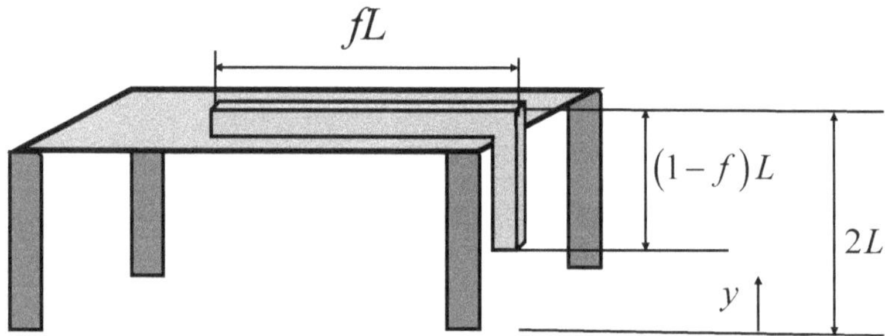

SOLUTION

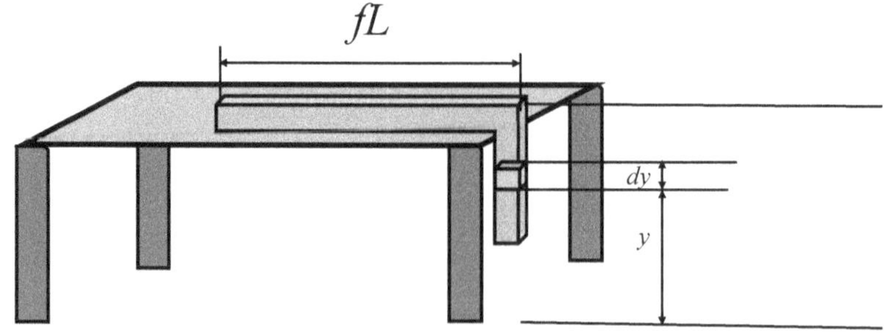

Electrostatic force on the hanging part is given by

$$F_1 = \int\limits_{2L-(1-f)L}^{2L} \frac{Q}{L} dy \frac{E_0}{y^2} \tag{1}$$

Simplifying

$$F_1 = \left[-\frac{Q}{L}\frac{E_0}{y} \right]_{2L-(1-f)L}^{2L}$$
$$\Rightarrow F_1 = \frac{QE_0(1-f)}{2L^2(1+f)} \tag{2}$$

Electrostatic force on the part resting on the table

$$F_2 = \frac{Q}{L}(fL)\frac{E_0}{(2L)^2}$$
$$\Rightarrow F_2 = \frac{QfE_0}{4L^2} \tag{3}$$

For vertical equilibrium of the resting part

$$N = F_2 \tag{4}$$

Static friction on the resting part must balance weight of the hanging part

$$fr_s = F_1 \tag{5}$$

Static friction must be less than or equal to its limiting value

$$fr_s \leq \mu N \tag{6}$$

Feeding equations 2-5 in equation 6

$$\frac{QE_0(1-f)}{2L^2(1+f)} \leq \mu \frac{QfE_0}{4L^2}$$
$$\Rightarrow \frac{1-f}{1+f} \leq \frac{\mu f}{2} \tag{7}$$

Rearranging

$$\frac{1-f}{1+f} - \frac{\mu f}{2} \leq 0$$

$$\Rightarrow \frac{2-2f-\mu f-\mu f^2}{2(1+f)} \leq 0 \tag{8}$$

$$\Rightarrow 2(1+f)(2-2f-\mu f-\mu f^2) \leq 0$$

$$\Rightarrow \mu f^2 + (2+\mu)f - 2 \geq 0$$

This is a quadratic inequality in f. The 2 roots of the corresponding equation are

$$\alpha = \frac{-(2+\mu) - \sqrt{(2+\mu)^2 + 8\mu}}{2\mu}$$

$$\tag{9}$$

$$\beta = \frac{-(2+\mu) + \sqrt{(2+\mu)^2 + 8\mu}}{2\mu}$$

The solution of the inequality is

$$f \in (-\infty, \alpha] \ \cup \ [\beta, \infty) \tag{10}$$

Since f must lie between 0 and 1, therefore

$$f \in [\beta, 1] \tag{11}$$

Hence,

$$f_{min} = \beta$$

$$\Rightarrow f_{min} = \frac{-(2+\mu) + \sqrt{(2+\mu)^2 + 8\mu}}{2\mu} \tag{12}$$

Question 18

A sphere of radius R floats in a dielectric liquid with two thirds of its height immersed. The sphere has a charge Q uniformly distributed over its volume. A point charge is fixed at a height h above the sphere as shown. Find the magnitude of the point charge for the sphere to be in equlibrium. Density of the liquid is ρ_f. The acceleration due to gravity on the Earth's surface is g and permittivity of free space is ε_0. Neglect weight of sphere.

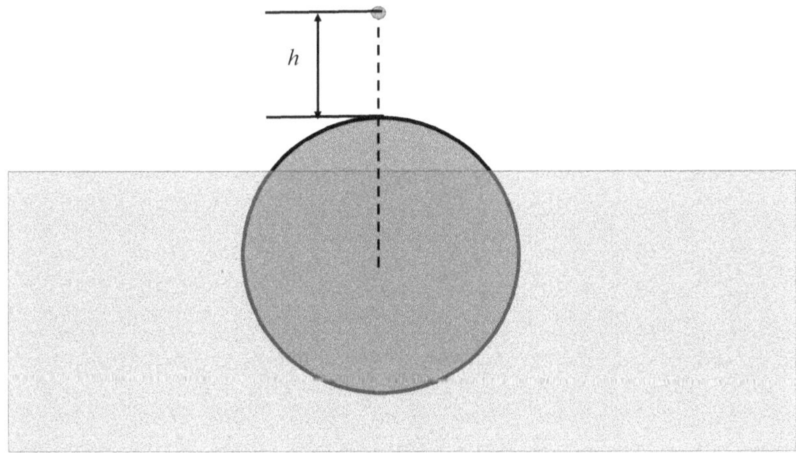

SOLUTION

For equilibrium of the sphere, the electrostatic force must balance the buoyant force

CALCULATION OF BOUYANT FORCE

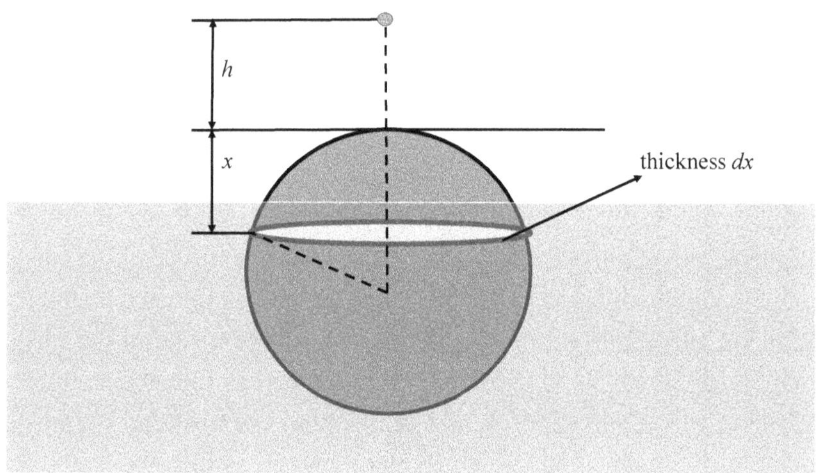

Calculation of bouyant force requires finding out the volume of sphere immersed. This is illustrated below

$$V_{in} = \int dV_{in} = \int_{\frac{2R}{3}}^{2R} \pi \left(\sqrt{R^2 - (R-x)^2} \right)^2 dx$$

$$\Rightarrow V_{in} = \pi \int_{\frac{2R}{3}}^{2R} \left(2Rx - x^2 \right) dx \tag{1}$$

$$V_{in} = \pi \left[Rx^2 - \frac{x^3}{3} \right]_{\frac{2R}{3}}^{2R}$$

$$\Rightarrow V_{in} = \frac{80\pi R^3}{81} \tag{2}$$

The buoyant force is equal to the weight of the fluid displaced

$$F_B = \frac{80\pi R^3 \rho_f g}{81} \tag{3}$$

CALCULATION OF ELECTROSTATIC FORCE

Since the submerged part of the sphere exists in a dielectric fluid, this part shall experience no electrostatic force. The net electrostatic force between the point charge and the sphere will therefore be the integral of the differential force between the shown disc and the point charge, with suitable limits that cover the entire part of the sphere outside the fluid.

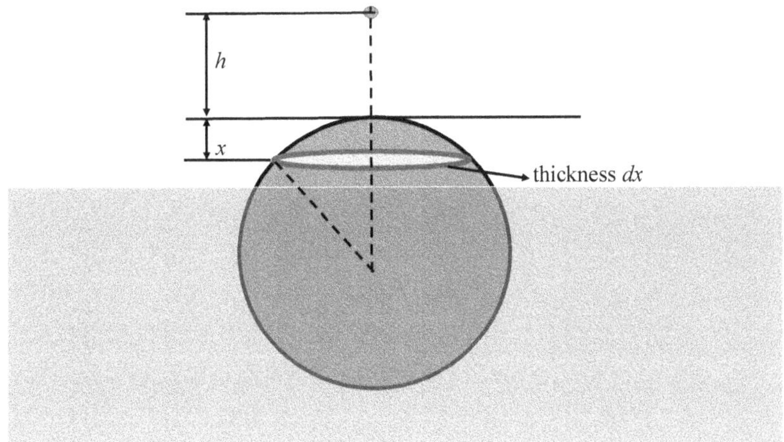

$$F = \int dF = \int_{0}^{\frac{2R}{3}} \frac{q\sigma}{2\varepsilon_0}\left[1 - \frac{x+h}{\sqrt{R^2-(R-x)^2}+(x+h)^2}\right] \qquad (4)$$

Here, σ is the charge per unit area on this disc. The charge on the disc will be charge per unit volume of sphere multiplied by the disc volume

$$\sigma = \frac{\dfrac{Q}{\frac{4}{3}\pi R^3}\pi\left[R^2-(R-x)^2\right]dx}{\pi\left[R^2-(R-x)^2\right]} - \frac{3Q}{4\pi R^3}dx \qquad (5)$$

Feeding equation 5 in 4

$$F = \frac{3Qq}{8\varepsilon_0 \pi R^3} \int_0^{\frac{2R}{3}} \left[1 - \frac{x+h}{\sqrt{R^2 - (R-x)^2 + (x+h)^2}} \right] dx \tag{6}$$

Simplifying

$$F = \frac{3Qq}{8\varepsilon_0 \pi R^3} \int_0^{\frac{2R}{3}} dx - \frac{3Qq}{8\varepsilon_0 \pi R^3} \int_0^{\frac{2R}{3}} \frac{x+h}{\sqrt{h^2 + 2x(h+R)}} dx \tag{7}$$

Let $h^2 + 2x(h+R) = U \Rightarrow x = \dfrac{U - h^2}{2(h+R)}$ and $dx = \dfrac{dU}{2(h+R)}$

$$F = \frac{Qq}{4\varepsilon_0 \pi R^2} - \frac{3Qq}{32\varepsilon_0 \pi R^3 (h+R)^2} \int_{h^2}^{h^2 + \frac{4R}{3}(h+R)} \frac{\left(U + h^2 + 2hR\right)}{\sqrt{U}} \cdot dU \tag{8}$$

Simplifying

$$F = \frac{Qq}{4\varepsilon_0 \pi R^2} - \frac{3Qq}{32\varepsilon_0 \pi R^3 (h+R)^2} \left[\int_{h^2}^{h^2 + \frac{4R}{3}(h+R)} \sqrt{U} \cdot dU + \int_{h^2}^{h^2 + \frac{4R}{3}(h+R)} \frac{\left(h^2 + 2hR\right) \cdot dU}{\sqrt{U}} \right] \tag{9}$$

$$F = \frac{Qq}{4\varepsilon_0 \pi R^2} - \frac{3Qq}{32\varepsilon_0 \pi R^3 (h+R)^2} \left[\left[\frac{2U^{\frac{3}{2}}}{3} \right]_{h^2}^{h^2 + \frac{4R}{3}(h+R)} + \left[2\left(h^2 + 2hR\right)\sqrt{U} \right]_{h^2}^{h^2 + \frac{4R}{3}(h+R)} \right] \tag{10}$$

$$F = \frac{Qq}{4\varepsilon_0 \pi R^2} - \frac{3Qq}{32\varepsilon_0 \pi R^3 (h+R)^2} \left[\frac{2\left\{ h^2 + \frac{4R}{3}(h+R) \right\}^{\frac{3}{2}}}{3} - \frac{2h^3}{3} + 2\left(h^2 + 2hR\right)\sqrt{h^2 + \frac{4R}{3}(h+R)} - 2\left(h^2 + 2hR\right)h \right] \tag{11}$$

Equating the buoyant and electrostatic forces

$$\frac{Qq}{4\varepsilon_0 \pi R^2} - \frac{3Qq}{32\varepsilon_0 \pi R^3 (h+R)^2} \left[\frac{2\left\{h^2 + \frac{4R}{3}(h+R)\right\}^{\frac{3}{2}}}{3} - \frac{2h^3}{3} + 2\left(h^2 + 2hR\right)\sqrt{h^2 + \frac{4R}{3}(h+R)} - 2\left(h^2 + 2hR\right)h \right] = \frac{80\pi R^3 \rho_f g}{81}$$

$$\Rightarrow q = \frac{\dfrac{320\pi^2 R^5 \varepsilon_0 \rho_f g}{81Q}}{1 - \dfrac{3\left[\dfrac{2\left\{h^2 + \frac{4R}{3}(h+R)\right\}^{\frac{3}{2}}}{3} - \dfrac{2h^3}{3} + 2\left(h^2 + 2hR\right)\sqrt{h^2 + \frac{4R}{3}(h+R)} - 2\left(h^2 + 2hR\right)h \right]}{8\varepsilon_0 R(h+R)^2}} \tag{12}$$

Question 19

A charged disc of radius R rotating at an angular velocity w_0 is placed on a rough horizontal surface at $t=0$. The friction coefficient for the disc surface pair is μ. Mass and charge are distributed on the disc unevenly as per the following functions: Mass per unit area $\sigma_m = ar^2$ and Charge per unit area $\sigma_q = ar^4$. A small loop of area A_0 is located at a height h above the disc along its axis, with its plane parallel to that of disc. The loop is small enough to be entirely located at this location. Find magnitude of instantaneous emf induced in the loop. The acceleration due to gravity on the Earth's surface is g and permeability of free space is μ_0. Neglect air drag and buoyancy.

SOLUTION

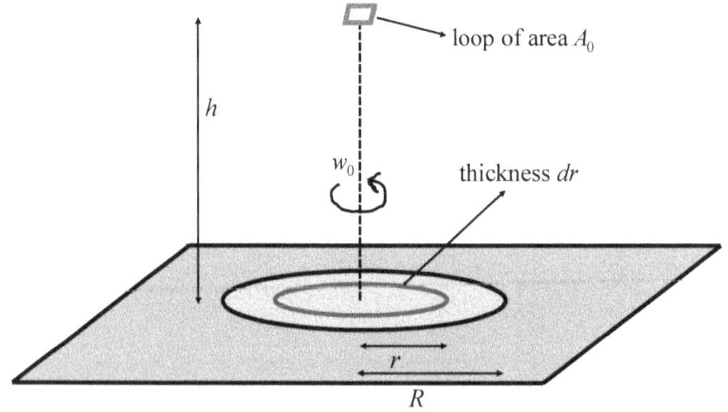

Friction between the disc and surface below will retard the disc and reduce its angular velocity. This reduces the disc's magnetic field at all locations around it, and therefore the magnetic field through the loop as well. At a general time t, when the disc's angular velocity has become w, its magnetic field at the loop is calculated as follows:

The magnetic field at the loop because of the differential ring shown is

$$dB = \frac{\mu_0 (di) r^2}{2\left(r^2 + h^2\right)^{\frac{3}{2}}} \tag{1}$$

The current flowing in this ring will be the charge on it divided by time period of its rotation

$$di = \frac{dq}{\dfrac{2\pi}{w}} = \frac{\sigma_q w\left(2\pi r \cdot dr\right)}{2\pi} = awr^5 \cdot dr \tag{2}$$

Feeding equation 2 in 1

$$dB = \frac{\mu_0 ar^7 w \cdot dr}{2\left(r^2 + h^2\right)^{\frac{3}{2}}} \tag{3}$$

The net magnetic field at the loop will be integral of above.

$$B = \int dB = \int_0^R \frac{\mu_0 awr^7 \cdot dr}{2\left(r^2 + h^2\right)^{\frac{3}{2}}} \tag{4}$$

Let $r^2 + h^2 = U^2 \Rightarrow r \cdot dr = U \cdot dU$

$$B = \frac{\mu_0 aw}{2} \int_h^{\sqrt{R^2 + h^2}} \frac{\left(U^2 - h^2\right)^3 dU}{U^2} \tag{5}$$

$$B = \frac{\mu_0 aw}{2} \int_h^{\sqrt{R^2+h^2}} \frac{\left(U^6 - h^6 - 3U^4 h^2 + 3U^2 h^4\right) dU}{U^2}$$

$$\Rightarrow B = \frac{\mu_0 aw}{2} \left[\int_h^{\sqrt{R^2+h^2}} U^4 dU - \int_h^{\sqrt{R^2+h^2}} \frac{h^6 \cdot dU}{U^2} - \int_h^{\sqrt{R^2+h^2}} 3U^2 h^2 \cdot dU + \int_h^{\sqrt{R^2+h^2}} 3h^4 \cdot dU \right] \tag{6}$$

$$B = \frac{\mu_0 aw}{2} \left[\frac{\left(R^2+h^2\right)^{\frac{5}{2}}}{5} - \frac{h^5}{5} + \frac{h^6}{\sqrt{R^2+h^2}} - h^5 - \left(R^2+h^2\right)^{\frac{3}{2}} h^2 + h^5 + 3h^4 \sqrt{R^2+h^2} - 3h^5 \right] \tag{7}$$

The magnetic flux through the loop at this instant will be

$$\phi = \frac{\mu_0 aw A_0}{2} \left[\frac{\left(R^2+h^2\right)^{\frac{5}{2}}}{5} - \frac{h^5}{5} + \frac{h^6}{\sqrt{R^2+h^2}} - h^5 - \left(R^2+h^2\right)^{\frac{3}{2}} h^2 + h^5 + 3h^4 \sqrt{R^2+h^2} - 3h^5 \right] \tag{8}$$

As per Faraday's law of electromagnetic induction, the emf induced in the loop at this instant will be

$$\left|V_{ind}\right| = \left|\frac{d\phi}{dt}\right|$$

$$\Rightarrow \left|V_{ind}\right| = \left| \frac{\mu_0 a A_0}{2} \frac{dw}{dt} \left[\frac{\left(R^2+h^2\right)^{\frac{5}{2}}}{5} - \frac{h^5}{5} + \frac{h^6}{\sqrt{R^2+h^2}} - h^5 - \left(R^2+h^2\right)^{\frac{3}{2}} h^2 + h^5 + 3h^4 \sqrt{R^2+h^2} - 3h^5 \right] \right| \tag{9}$$

To find the angular velocity of disc as a function of time, torque on the disc needs to be found out and then rotational analogue of Newton's second law of motion needs to be used. Fictional torque acting on the differential disc shown is

$$d\tau = \mu \cdot dN \cdot r = \mu\left(dm\right)gr = \mu\left(\sigma_m \cdot 2\pi r \cdot dr\right)gr \tag{10}$$

The net frictional torque on the disc is the integral of above

$$\tau = \int d\tau = \int_0^R -2\pi\mu agr^4 \cdot dr = -\frac{2\pi\mu agR^5}{5} \tag{11}$$

The moment of inertia of disc about the relevant axis is

$$I = \int dm \cdot r^2 = \int \sigma_m \left(2\pi r \cdot dr \right) r^2 = \int_0^R 2\pi a r^5 \cdot dr = \frac{\pi a R^6}{3} \tag{12}$$

Applying rotational analogue of Newton's second law of motion

$$\tau = I \frac{dw}{dt}$$
$$\Rightarrow \frac{dw}{dt} = -\frac{6\mu g}{5R} \tag{13}$$

Feeding equation 13 in equation 9

$$\left| V_{ind} \right| = \frac{3\mu_0 a A_0 \mu g}{5R} \left[\frac{\left(R^2 + h^2 \right)^{\frac{5}{2}}}{5} - \frac{h^5}{5} + \frac{h^6}{\sqrt{R^2 + h^2}} - h^5 - \left(R^2 + h^2 \right)^{\frac{3}{2}} h^2 + h^5 + 3h^4 \sqrt{R^2 + h^2} - 3h^5 \right] \tag{14}$$

Question 20

An alternator or an AC dynamo or an AC generator works as follows: A loop made of an electrically conducting material is placed in a constant magnetic field. By some force, it is rotated in this magnetic field at a constant angular velocity. As it rotates, the angle between the magnetic field and the loop's area vector changes, and so does magnetic flux across the loop. Due to changing magnetic flux, an emf is induced in the loop. Consider this setup. Suppose the loop isn't continuously rotated but just imparted an initial angular velocity w_0. Air resistance gradually reduces the angular velocity. If the loop contains an electrical resistor R, find total heat generated in the resistor until the loop comes to rest. The moment of inertia of the loop about the rotation axis is I. The loop's cross-sectional area is A, and magnetic field it is placed in is B. Air resistance torque on the loop at any instant=kw, where k is a positive constant.

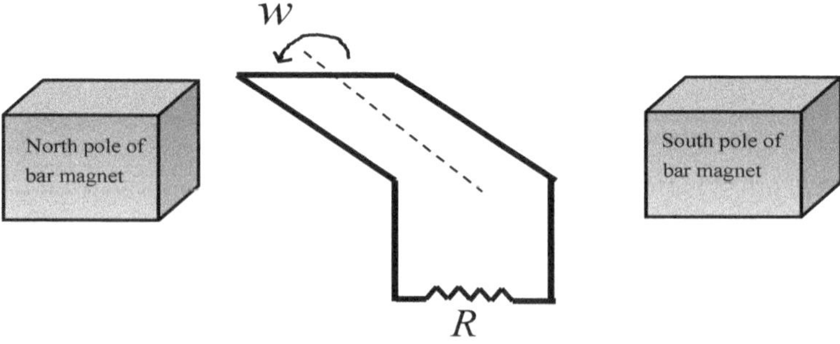

SOLUTION

Applying rotational analogue of Newton's second law of motion on the loop at a general instant

$$-kw = I\frac{dw}{dt} \tag{1}$$

Separating the variables and integrating both sides

$$\int_{w_0}^{w}\frac{dw}{w} = -\int_{0}^{t}\frac{k}{I}dt \tag{2}$$

$$\Rightarrow w = w_0 e^{-\frac{kt}{I}}$$

Using the differential definition of angular velocity $w = \dfrac{d\theta}{dt}$, separating the variables and integrating both sides

$$\int_{0}^{\theta}d\theta = \int_{0}^{t}w_0 e^{-\frac{kt}{I}}dt \tag{3}$$

$$\Rightarrow \theta = \frac{Iw_0}{k}\left(1 - e^{-\frac{kt}{I}}\right)$$

The instantaneous magnetic flux through the loop is given by

$$\phi = BA\cos\theta \tag{4}$$

As per faraday's law of electromagnetic induction, the magnitude of the instantaneous emf induced in the loop will be

$$|V| = \left|\frac{d\phi}{dt}\right| = BA\sin\left[\frac{Iw_0}{k}\left(1 - e^{-\frac{kt}{I}}\right)\right]w_0 e^{-\frac{kt}{I}} \tag{5}$$

There is only one electrical resistance in the circuit. So, the instantaneous power generated is

$$P = \frac{V^2}{R} = \frac{B^2 A^2 w_0^2 e^{-\frac{2kt}{I}}}{R} \sin^2\left[\frac{Iw_0}{k}\left(1 - e^{-\frac{kt}{I}}\right)\right] \tag{6}$$

Heat will continue to be generated from the resistor until current is flowing. And this shall go on until there is a voltage in the loop. Equation 5 tells that this shall occur until $t \to \infty$. So total heat generated will be integral of the expression of power with respect to time from $t=0$ to $t \to \infty$

$$\text{Heat generated} = \int_0^\infty P \cdot dt$$

$$\Rightarrow \text{Heat generated} = \frac{B^2 A^2 w_0^2}{R} \int_0^\infty \sin^2\left[\frac{Iw_0}{k}\left(1 - e^{-\frac{kt}{I}}\right)\right] e^{-\frac{2kt}{I}} \cdot dt \tag{7}$$

Let $\dfrac{Iw_0}{k}\left(1 - e^{-\frac{kt}{I}}\right) = U \Rightarrow w_0 e^{-\frac{kt}{I}} dt = dU$ or $dt = \dfrac{dU}{w_0\left(1 - \dfrac{kU}{Iw_0}\right)} = \dfrac{I \cdot dU}{Iw_0 - kU}$

$$\text{Heat generated} = \frac{B^2 A^2 w_0^2}{R} \int_0^{\frac{Iw_0}{k}} \sin^2 U \left(1 - \frac{kU}{Iw_0}\right)^2 \cdot \frac{I \cdot dU}{Iw_0 - kU} \tag{8}$$

Simplifying

$$\text{Heat generated} = \frac{B^2 A^2}{IR} \int_0^{\frac{Iw_0}{k}} \left(\sin^2 U\right)\left(Iw_0 - kU\right) dU \tag{9}$$

$$\text{Heat generated} = \frac{B^2 A^2}{2IR} \int_0^{\frac{Iw_0}{k}} \left(1 - \cos 2U\right)\left(Iw_0 - kU\right) dU \tag{10}$$

Using integration by parts

$$\text{Heat generated} = \frac{B^2 A^2}{2IR}\left[\frac{k\cos 2U}{4} - \frac{\left(Iw_0 - kU\right)\sin 2U}{2}\right]_0^{\frac{Iw_0}{k}} \tag{11}$$

$$\text{Heat generated} = \frac{B^2 A^2 k}{8IR}\left[\cos\left(\frac{2Iw_0}{k}\right) - 1\right] \tag{12}$$

Question 21

Consider the previous question. If air drag is neglected, and the magnetic field produced by the bar magnet decays with time as per the relation: $B = B_0 e^{-\alpha t}$ (α is a positive constant), find heat generated upto time t

SOLUTION

Since there is no air drag, and no other force acts on the loop, it will continue to rotate at its initial angular velocity w_0. The angle rotated at any time t is therefore

$$\theta = w_0 t \tag{1}$$

The instantaneous magnetic flux through the loop is

$$\phi = BA\cos\theta = B_0 e^{-\alpha t} A\cos\left(w_0 t\right) \tag{2}$$

As per Faraday's law of electromagnetic induction, the magnitude of instantaneous induced emf in the loop is

$$|V| = \left|\frac{d\phi}{dt}\right| = B_0 A e^{-\alpha t}\left[w_0 \sin\left(w_0 t\right) + \alpha\cos\left(w_0 t\right)\right] \tag{3}$$

The instantaneous power generated across the electrical resistor is

$$P = \frac{V^2}{R} \tag{4}$$

The heat generated upto time t will be

$$\text{Heat Generated} = \int_0^t P \cdot dt \tag{5}$$

Feeding equations 3 and 4 in 5

$$\text{Heat Generated} = \int_0^t \frac{B_0^2 A^2 e^{-2\alpha t}\left[w_0^2 \sin^2\left(w_0 t\right) + \alpha^2 \cos^2\left(w_0 t\right) + \alpha w_0 \sin\left(2w_0 t\right)\right]}{R} \cdot dt \tag{6}$$

Simplifying

$$\text{Heat Generated} = \frac{B_0^2 A^2}{R} \int_0^t e^{-2\alpha t}\left[w_0^2 \left\{\frac{1-\cos\left(2w_0 t\right)}{2}\right\} + \alpha^2 \left\{\frac{1+\cos\left(2w_0 t\right)}{2}\right\} + \alpha w_0 \sin\left(2w_0 t\right)\right] \cdot dt \tag{7}$$

$$\text{Heat Generated} = \frac{B_0^2 A^2}{R}\left[\int_0^t \left(\frac{w_0^2 + \alpha^2}{2}\right) e^{-2\alpha t} \cdot dt + \frac{\left(\alpha^2 - w_0^2\right)}{2} \int_0^t e^{-2\alpha t}\cos\left(2w_0 t\right) \cdot dt + \alpha w_0 \int_0^t e^{-2\alpha t}\sin\left(2w_0 t\right) \cdot dt \right] \tag{8}$$

$$\text{Heat Generated} = \frac{B_0^2 A^2}{R}\left[\begin{array}{c} \dfrac{\left(w_0^2 + \alpha^2\right)\left(1 - e^{-2\alpha t}\right)}{4\alpha} + \\[2ex] \dfrac{\left(\alpha^2 - w_0^2\right)}{2} \displaystyle\int_0^t e^{-2\alpha t}\cos\left(2w_0 t\right) \cdot dt + \\[2ex] \alpha w_0 \displaystyle\int_0^t e^{-2\alpha t}\sin\left(2w_0 t\right) \cdot dt \end{array} \right] \tag{9}$$

The two integrals appearing in equation 9 have to be evaluated by integration by parts. Let them respectively be denoted by Int_1 and Int_2. Their evaluation is shown below

$$Int_1 = \cos(2w_0 t)\frac{e^{-2\alpha t}}{-2\alpha} - \int -2w_0 \sin(2w_0 t)\frac{e^{-2\alpha t}}{-2\alpha}\,dt$$

$$\Rightarrow Int_1 = -\cos(2w_0 t)\frac{e^{-2\alpha t}}{2\alpha} - \frac{w_0}{\alpha}\int \sin(2w_0 t)e^{-2\alpha t}\,dt$$

$$\Rightarrow Int_1 = -\cos(2w_0 t)\frac{e^{-2\alpha t}}{2\alpha} - \frac{w_0}{\alpha}\left[\sin(2w_0 t)\frac{e^{-2\alpha t}}{-2\alpha} - \int 2w_0 \cos(2w_0 t)\frac{e^{-2\alpha t}}{-2\alpha}\,dt\right]$$

$$\Rightarrow Int_1 = -\cos(2w_0 t)\frac{e^{-2\alpha t}}{2\alpha} - \frac{w_0}{\alpha}\left[\sin(2w_0 t)\frac{e^{-2\alpha t}}{-2\alpha} + \frac{w_0}{\alpha}Int_1\right] \qquad (10)$$

$$\Rightarrow Int_1 = -\cos(2w_0 t)\frac{e^{-2\alpha t}}{2\alpha} + \frac{w_0}{2\alpha^2}\sin(2w_0 t)e^{-2\alpha t} - \frac{w_0^2}{\alpha^2}Int_1$$

$$\Rightarrow Int_1\left[1 + \frac{w_0^2}{\alpha^2}\right] = -\cos(2w_0 t)\frac{e^{-2\alpha t}}{2\alpha} + \frac{w_0}{2\alpha^2}\sin(2w_0 t)e^{-2\alpha t}$$

$$\Rightarrow Int_1 = \frac{e^{-2\alpha t}}{2}\left[\frac{w_0 \sin(2w_0 t) - \alpha \cos(2w_0 t)}{w_0^2 + \alpha^2}\right]$$

$$Int_2 = \sin(2w_0 t)\frac{e^{-2\alpha t}}{-2\alpha} - \int 2w_0 \cos(2w_0 t)\frac{e^{-2\alpha t}}{-2\alpha}\,dt$$

$$\Rightarrow Int_2 = -\sin(2w_0 t)\frac{e^{-2\alpha t}}{2\alpha} + \frac{w_0}{\alpha}\int \cos(2w_0 t)e^{-2\alpha t}\,dt$$

$$\Rightarrow Int_2 = -\sin(2w_0 t)\frac{e^{-2\alpha t}}{2\alpha} + \frac{w_0}{\alpha}\left[\cos(2w_0 t)\frac{e^{-2\alpha t}}{-2\alpha} - \int -2w_0 \sin(2w_0 t)\frac{e^{-2\alpha t}}{-2\alpha}\,dt\right]$$

$$\Rightarrow Int_2 = -\sin(2w_0 t)\frac{e^{-2\alpha t}}{2\alpha} + \frac{w_0}{\alpha}\left[\cos(2w_0 t)\frac{e^{-2\alpha t}}{-2\alpha} - \frac{w_0}{\alpha}Int_2\right] \qquad (11)$$

$$\Rightarrow Int_2 = -\sin(2w_0 t)\frac{e^{-2\alpha t}}{2\alpha} - \frac{w_0}{2\alpha^2}\cos(2w_0 t)e^{-2\alpha t} - \frac{w_0^2}{\alpha^2}Int_2$$

$$\Rightarrow Int_2\left[1 + \frac{w_0^2}{\alpha^2}\right] = -\sin(2w_0 t)\frac{e^{-2\alpha t}}{2\alpha} - \frac{w_0}{2\alpha^2}\cos(2w_0 t)e^{-2\alpha t}$$

$$\Rightarrow Int_2 = -\frac{e^{-2\alpha t}}{2}\left[\frac{\alpha \sin(2w_0 t) + w_0 \cos(2w_0 t)}{\alpha^2 + w_0^2}\right]$$

Feeding equations 10 and 11 in equation 9

$$\text{Heat Generated} = \frac{B_0^2 A^2}{R}\left[\frac{\left(w_0^2+\alpha^2\right)\left(1-e^{-2\alpha t}\right)}{4\alpha}+\frac{\left(\alpha^2-w_0^2\right)}{2}\left[\frac{e^{-2\alpha t}w_0\sin\left(2w_0t\right)-e^{-2\alpha t}\alpha\cos\left(2w_0t\right)+\alpha}{2\left(w_0^2+\alpha^2\right)}\right]+\alpha w_0\left[\frac{w_0-e^{-2\alpha t}\alpha\sin\left(2w_0t\right)-e^{-2\alpha t}w_0\cos\left(2w_0t\right)}{2\left(\alpha^2+w_0^2\right)}\right]\right] \qquad (12)$$

Question 22

At $t=0$, the shown semicircular loop begins to rotate in the XY plane about an axis passing through its centre and in the third dimension (into the plane of paper). The rotation occurs at a constant angular velocity w. A magnetic field exists in the region into the plane of paper, with its magnitude directly proportional to distance from the Y axis, the proportionality constant being a. Find instantaneous emf induced in the loop.

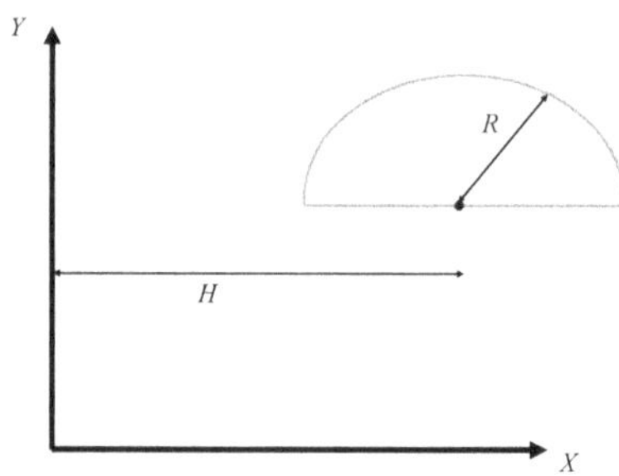

SOLUTION

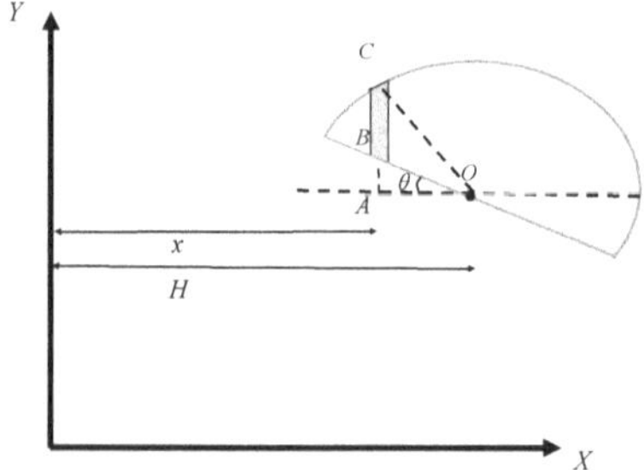

The magnetic flux through the loop at a general time t is found by integrating the differential flux through the shown strip across suitable limits

$$d\phi = B_x\left(\text{BC}\right)\left(dx\right)$$
$$\Rightarrow d\phi = B_x\left(\text{AC-AB}\right)\left(dx\right) \tag{1}$$

Using geometry

$$\frac{AB}{OA} = \tan\theta$$
$$\Rightarrow AB = \left(H-x\right)\tan wt \tag{2}$$

$$\left(\text{OA}\right)^2 + \left(\text{AC}\right)^2 = \left(\text{OC}\right)^2$$
$$\Rightarrow \text{AC} = \sqrt{R^2 - \left(H-x\right)^2} \tag{3}$$

Feeding equations 2 and 3 in 1

$$d\phi = ax\left[\sqrt{R^2 - \left(H-x\right)^2} - \left(H-x\right)\tan wt\right]\cdot dx \tag{4}$$

The net flux is

$$\phi = \int d\phi$$
$$\Rightarrow \phi = \int_{H-R\cos(wt)}^{H+R} ax\left[\sqrt{R^2 - \left(H-x\right)^2} - \left(H-x\right)\tan wt\right]\cdot dx \tag{5}$$

Simplifying

$$\phi = \left[a\int_{H-R\cos(wt)}^{H+R} x\sqrt{R^2 - \left(H-x\right)^2}\cdot dx\right] - \left[a\int_{H-R\cos(wt)}^{H+R} \left(Hx - x^2\right)\tan\left(wt\right)\cdot dx\right] \tag{6}$$

Let $H - x = R\sin\beta \Rightarrow x = H - R\sin\beta$ and $dx = -R\cos\beta\cdot d\beta$

$$\phi = \left[a \int_{\sin^{-1}\cos wt}^{\sin^{-1}-1} (H - R\sin\beta)(R\cos\beta)\cdot(-R\cos\beta)d\beta \right] - \left[a \int_{H-R\cos(wt)}^{H+R} \left(Hx - x^2\right)\tan(wt)\cdot dx \right]$$

$$\Rightarrow \phi = -\left[aR^2 H \int_{\sin^{-1}\cos wt}^{\sin^{-1}-1} \cos^2\beta\cdot d\beta \right] + \left[aR^3 \int_{\sin^{-1}\cos wt}^{\sin^{-1}-1} \sin\beta\cos^2\beta\cdot d\beta \right] - \left[a\tan(wt) \int_{H-R\cos(wt)}^{H+R} \left(Hx - x^2\right)\cdot dx \right] \quad (7)$$

$$\phi = -\left[aR^2 H \int_{\sin^{-1}\cos wt}^{\sin^{-1}-1} \left(\frac{1+\cos 2\beta}{2}\right)\cdot d\beta \right] - \left[aR^3 \int_{\sin^{-1}\cos wt}^{\sin^{-1}-1} \cos^2\beta\cdot d(\cos\beta) \right] - \left[a\tan(wt) \int_{H-R\cos(wt)}^{H+R} \left(Hx - x^2\right)\cdot dx \right] \quad (8)$$

$$\phi = -\frac{aR^2 H}{2}\left[\beta + \frac{\sin 2\beta}{2} \right]_{\sin^{-1}\cos wt}^{-\frac{\pi}{2}} - aR^3\left[\frac{\cos^3\beta}{3} \right]_{\sin^{-1}\cos wt}^{-\frac{\pi}{2}} - a\tan(wt)\left[\frac{Hx^2}{2} - \frac{x^3}{3} \right]_{H-R\cos(wt)}^{H+R} \quad (9)$$

$$\phi = -\frac{aR^2 H\left[-\frac{\pi}{2} - \sin^{-1}\cos wt - \frac{\sin 2wt}{2} \right]}{2} + $$
$$\frac{aR^3 \sin^3 wt}{3} - $$
$$a\tan wt\left[\frac{H(H+R)^2}{2} - \frac{(H+R)^3}{3} - \frac{H(H-R\cos wt)^2}{2} + \frac{(H-R\cos wt)^3}{3} \right] \quad (10)$$

As per faraday's law of electromagnetic induction, the instantaneous emf induced in the loop will be

$$\left| V_{ind} \right| = \left| \frac{d\phi}{dt} \right|$$

$$\Rightarrow \left| V_{ind} \right| = -\frac{aR^2 Hw(1-\cos 2wt)}{2} + $$
$$awR^3 \sin^2 wt\cos wt + $$
$$\frac{Haw(H - R\cos wt)\left[H\sec^2 wt - R\sec wt + 2R\tan wt\sin wt \right]}{2} - $$
$$\frac{aw(H - R\cos wt)^2\left[H\sec^2 wt - R\sec wt + 3R\tan wt\sin wt \right]}{3} \quad (11)$$

Question 23

A wire is wound over a hollow hyperboloid as shown. A time dependent current $I_0 e^{-kt}$ flows through the wire (I_0 and k are positive constants). Permeability of the free space is μ_0, number of turns of the wire per unit length (length measured along Y axis shown) is n. A small loop of cross-sectional area A_0 is fixed at the origin, with its axis along the Y axis. The loop is small enough to be approximated entirely at the origin. Find magnitude of the instantaneous emf induced in the loop.

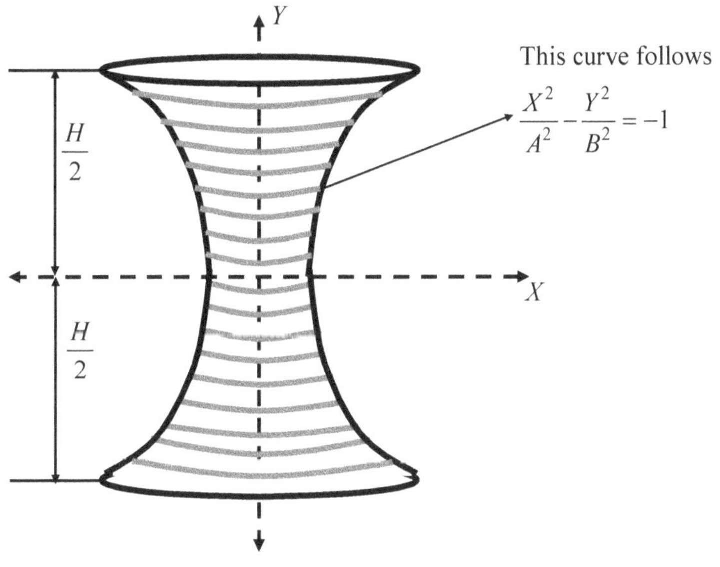

SOLUTION

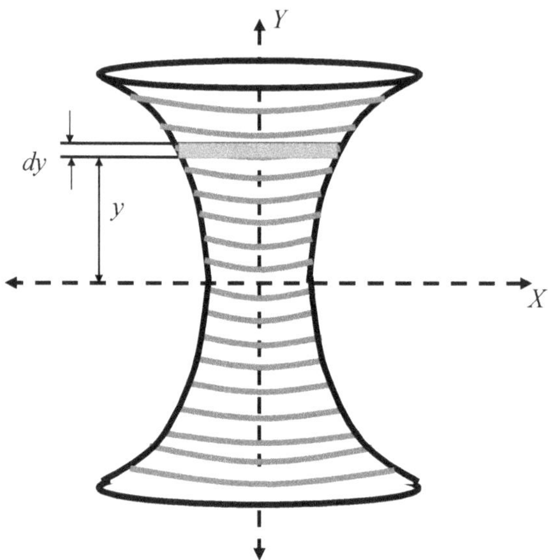

At a general time *t*, when the current passing through the wire is *I*, the magnetic field at the origin due to the differential cylinder shown is

$$dF_m = \frac{\mu_0 \left(n \cdot dy \right) I r^2}{2 \left(r^2 + y^2 \right)^{\frac{3}{2}}} \tag{1}$$

The net magnetic field at the origin shall be integral of above, with limits covering the entire hyperboloid

$$F_m = \int dF_m = \frac{\mu_0 n I}{2} \int \frac{r^2 dy}{\left(r^2 + y^2 \right)^{\frac{3}{2}}} \tag{2}$$

r is a function of *y*. So this functional dependence needs to be sought in order to evaluate the above integral. All points with coordinates (*r,y*)will satisfy the given curve equation. Therefore,

$$r^2 = \frac{A^2}{B^2}\left(y^2 - B^2\right) \tag{3}$$

Feeding equation 3 in equation 2

$$F_m = \frac{\mu_0 nIA^2 B}{2} \int_{-\frac{H}{2}}^{\frac{H}{2}} \frac{\left(y^2 - B^2\right)dy}{\left[\left(A^2 + B^2\right)y^2 - A^2 B^2\right]^{\frac{3}{2}}} \tag{4}$$

Simplifying

$$F_m = \frac{\mu_0 nIA^2 B}{2\left(A^2 + B^2\right)^{\frac{3}{2}}} \int_{-\frac{H}{2}}^{\frac{H}{2}} \frac{\left(y^2 - B^2\right)dy}{\left[y^2 - \frac{A^2 B^2}{A^2 + B^2}\right]^{\frac{3}{2}}} \tag{5}$$

Let $y = \dfrac{AB}{\sqrt{A^2 + B^2}}\sec\theta \Rightarrow dy = \dfrac{AB}{\sqrt{A^2 + B^2}}\sec\theta\tan\theta \cdot d\theta$

$$F_m = \frac{\mu_0 nIA^2 B}{2\left(A^2 + B^2\right)^{\frac{3}{2}}} \int_{\sec^{-1}\left(\frac{H}{2}\frac{\sqrt{A^2+B^2}}{AB}\right)}^{\sec^{-1}\left(\frac{H}{2}\frac{\sqrt{A^2+B^2}}{AB}\right)} \frac{\left(\frac{A^2 B^2 \sec^2\theta}{A^2 + B^2} - B^2\right)\frac{AB}{\sqrt{A^2 + B^2}}\sec\theta\tan\theta \cdot d\theta}{\left(\frac{A^2 B^2 \sec^2\theta}{A^2 + B^2} - \frac{A^2 B^2}{A^2 + B^2}\right)^{\frac{3}{2}}} \tag{6}$$

$$F_m = \frac{\mu_0 nIB}{2\sqrt{A^2 + B^2}} \int_{\sec^{-1}\left(-\frac{H}{2}\frac{\sqrt{A^2+B^2}}{AB}\right)}^{\sec^{-1}\left(\frac{H}{2}\frac{\sqrt{A^2+B^2}}{AB}\right)} \frac{\left(\frac{A^2 \sec^2\theta}{A^2 + B^2} - 1\right)\cos\theta \cdot d\theta}{\sin^2\theta} \tag{7}$$

$$F_m = \frac{\mu_0 n I A^2 B}{2\left(A^2 + B^2\right)^{\frac{3}{2}}} \int\limits_{\sec^{-1}\left(-\frac{H}{2}\frac{\sqrt{A^2+B^2}}{AB}\right)}^{\sec^{-1}\left(\frac{H}{2}\frac{\sqrt{A^2+B^2}}{AB}\right)} \frac{\sec\theta}{\sin^2\theta}\cdot d\theta$$

$$-\frac{\mu_0 n I B}{2\sqrt{A^2+B^2}} \int\limits_{\sec^{-1}\left(-\frac{H}{2}\frac{\sqrt{A^2+B^2}}{AB}\right)}^{\sec^{-1}\left(\frac{H}{2}\frac{\sqrt{A^2+B^2}}{AB}\right)} \frac{\cos\theta}{\sin^2\theta}\cdot d\theta \tag{8}$$

$$F_m = \frac{\mu_0 n I A^2 B}{2\left(A^2 + B^2\right)^{\frac{3}{2}}} \int\limits_{\sec^{-1}\left(-\frac{H}{2}\frac{\sqrt{A^2+B^2}}{AB}\right)}^{\sec^{-1}\left(\frac{H}{2}\frac{\sqrt{A^2+B^2}}{AB}\right)} \frac{d\left(\sin\theta\right)}{\left(1-\sin^2\theta\right)\sin^2\theta}$$

$$-\frac{\mu_0 n I B}{2\sqrt{A^2+B^2}} \int\limits_{\sec^{-1}\left(-\frac{H}{2}\frac{\sqrt{A^2+B^2}}{AB}\right)}^{\sec^{-1}\left(\frac{H}{2}\frac{\sqrt{A^2+B^2}}{AB}\right)} \frac{d\left(\sin\theta\right)}{\sin^2\theta} \tag{9}$$

$$F_m = \frac{\mu_0 n I A^2 B}{2\left(A^2 + B^2\right)^{\frac{3}{2}}} \int\limits_{\sec^{-1}\left(-\frac{H}{2}\frac{\sqrt{A^2+B^2}}{AB}\right)}^{\sec^{-1}\left(\frac{H}{2}\frac{\sqrt{A^2+B^2}}{AB}\right)} \left[\frac{1}{1-\sin^2\theta} + \frac{1}{\sin^2\theta}\right] d\left(\sin\theta\right)$$

$$-\frac{\mu_0 n I B}{2\sqrt{A^2+B^2}} \int\limits_{\sec^{-1}\left(-\frac{H}{2}\frac{\sqrt{A^2+B^2}}{AB}\right)}^{\sec^{-1}\left(\frac{H}{2}\frac{\sqrt{A^2+B^2}}{AB}\right)} \frac{d\left(\sin\theta\right)}{\sin^2\theta} \tag{10}$$

$$F_m = \frac{\mu_0 n I A^2 B}{2\left(A^2+B^2\right)^{\frac{3}{2}}} \int\limits_{\sec^{-1}\left(-\frac{H}{2}\frac{\sqrt{A^2+B^2}}{AB}\right)}^{\sec^{-1}\left(\frac{H}{2}\frac{\sqrt{A^2+B^2}}{AB}\right)} \frac{d\left(\sin\theta\right)}{1-\sin^2\theta} -$$

$$\frac{\mu_0 n I B^3}{2\left(A^2+B^2\right)^{\frac{3}{2}}} \int\limits_{\sec^{-1}\left(-\frac{H}{2}\frac{\sqrt{A^2+B^2}}{AB}\right)}^{\sec^{-1}\left(\frac{H}{2}\frac{\sqrt{A^2+B^2}}{AB}\right)} \frac{d\left(\sin\theta\right)}{\sin^2\theta} \tag{11}$$

$$F_m = -\frac{\mu_0 n I A^2 B}{4\left(A^2+B^2\right)^{\frac{3}{2}}}\left[\ln\left|\frac{\sin\theta-1}{\sin\theta+1}\right|\right]_{\sec^{-1}\left(-\frac{H}{2}\frac{\sqrt{A^2+B^2}}{AB}\right)}^{\sec^{-1}\left(\frac{H}{2}\frac{\sqrt{A^2+B^2}}{AB}\right)} +$$

$$\frac{\mu_0 n I B^3}{2\left(A^2+B^2\right)^{\frac{3}{2}}}\left[\frac{1}{\sin\theta}\right]_{\sec^{-1}\left(-\frac{H}{2}\frac{\sqrt{A^2+B^2}}{AB}\right)}^{\sec^{-1}\left(\frac{H}{2}\frac{\sqrt{A^2+B^2}}{AB}\right)} \tag{12}$$

$$F_m = \frac{\mu_0 n I A^2 B}{2\left(A^2+B^2\right)^{\frac{3}{2}}} \ln\left|\frac{\sqrt{H^2\left(A^2+B^2\right)-4A^2 B^2}+H\sqrt{A^2+B^2}}{\sqrt{H^2\left(A^2+B^2\right)-4A^2 B^2}-H\sqrt{A^2+B^2}}\right| -$$

$$\frac{\mu_0 n I B^3}{2\left(A^2+B^2\right)^{\frac{3}{2}}} \tag{13}$$

As per Faraday's law of electromagnetic induction, the instantaneous emf induced in the loop will be

$$\left| V_{ind} \right| = \left| \frac{d\phi}{dt} \right|$$

$$\Rightarrow \left| V_{ind} \right| = A_0 \left| \frac{dF_m}{dt} \right| \tag{14}$$

Feeding equation 13 in equation 14

$$\left| V_{ind} \right| = \frac{\mu_0 n B A_0 I_0 k e^{-kt}}{2\left(A^2 + B^2\right)^{\frac{3}{2}}} \left[A^2 \ln \left| \frac{\sqrt{H^2\left(A^2 + B^2\right) - 4A^2 B^2} + H\sqrt{A^2 + B^2}}{\sqrt{H^2\left(A^2 + B^2\right) - 4A^2 B^2} - H\sqrt{A^2 + B^2}} \right| - B^2 \right] \tag{15}$$

Question 24

❖ ❖ ❖

Aparticle of mass m and charge q rests at the origin. A small loop (area A) made of a conducting material is located on the y axis at a distance h from the origin [The loop is small enough to be assumed to exist entirely at this coordinate]. At t=0, the particle is imparted a velocity v_0 along the positive x axis. **U**nder the action of various forces, it moves along the x axis in a manner that produces a constant induced emf V in the loop. Find the particle's displacement as a time function. Permeability of free space is μ_0

SOLUTION

As per the faraday's law of electromagnetic induction

$$V = -\frac{d\phi}{dt} \tag{1}$$

Separating the variables and integrating both sides

$$\int_{\phi_0}^{\phi} d\phi = -\int_0^t V \cdot dt \tag{2}$$

$$\Rightarrow \phi = \phi_0 - Vt$$

By definition, the instantaneous magnetic flux through the loop is

$$\phi = \vec{B} \cdot \vec{A}$$

$$\Rightarrow \phi = \frac{\mu_0 q \vec{v} \times \vec{r}}{4\pi |\vec{r}|^3} \cdot \vec{A} \tag{3}$$

Where $\vec{r}$ is the vector starting from the charge and ending at the location where field is needed. It is the difference of the position vectors of the latter and former. Here, $\vec{r} = \left(0\hat{i} + h\hat{j}\right) - \left(x\hat{i} + 0\hat{j}\right) = -x\hat{i} + h\hat{j}$

$$\phi = \frac{\mu_0 q \left(\dfrac{dx}{dt}\hat{i}\right) \times \left(-x\hat{i} + h\hat{j}\right)}{4\pi \left(x^2 + h^2\right)^{\frac{3}{2}}} \cdot A\hat{k} \tag{4}$$

$$\phi = \frac{\mu_0 q h \left(\dfrac{dx}{dt}\hat{k}\right)}{4\pi \left(x^2 + h^2\right)^{\frac{3}{2}}} \cdot A\hat{k} \tag{5}$$

$$\Rightarrow \phi = \frac{\mu_0 q h A \dfrac{dx}{dt}}{4\pi \left(x^2 + h^2\right)^{\frac{3}{2}}} \quad \text{and} \quad \phi_0 = \frac{\mu_0 q A v_0}{4\pi h^2}$$

Using equations 2 and 5

$$\frac{\mu_0 q h A \dfrac{dx}{dt}}{4\pi \left(x^2 + h^2\right)^{\frac{3}{2}}} = \frac{\mu_0 q A v_0}{4\pi h^2} - Vt \tag{6}$$

Separating the variables and integrating both sides

$$\frac{\mu_0 q h A}{4\pi}\int_0^x \frac{dx}{\left(x^2+h^2\right)^{\frac{3}{2}}} = \int_0^t\left(\frac{\mu_0 q A v_0}{4\pi h^2}-Vt\right)dt \tag{7}$$

Let $x = h\tan\theta \Rightarrow dx = h\sec^2\theta\cdot d\theta$

$$\frac{\mu_0 q A}{4\pi h}\int_0^{\tan^{-1}\frac{x}{h}} \cos\theta\cdot d\theta = \int_0^t\left(\frac{\mu_0 q A v_0}{4\pi h^2}-Vt\right)dt \tag{8}$$

$$\Rightarrow \frac{\mu_0 q A x}{4\pi h\sqrt{x^2+h^2}} = \frac{\mu_0 q A v_0 t}{4\pi h^2}-\frac{Vt^2}{2}$$

Rearranging

$$x = \frac{\mu_0 q A v_0 t - 2\pi V h^2 t^2}{\sqrt{\mu_0^2 q^2 A^2 - \left(\mu_0 q A v_0 t - 2\pi h^2 V t^2\right)^2}} \tag{9}$$

Question 25

An arbitrary shaped wire carrying a current I lies in the XY plane as shown. A magnetic field $\vec{B} = B_1\hat{i} + B_2\hat{j}$ exists in the region (B_1 and B_2 are constants). The magnetic force on the segment of the wire extending from origin to any x coordinate is proportional to the slope of the tangent at that point (proportionality constant being a). Find the function f

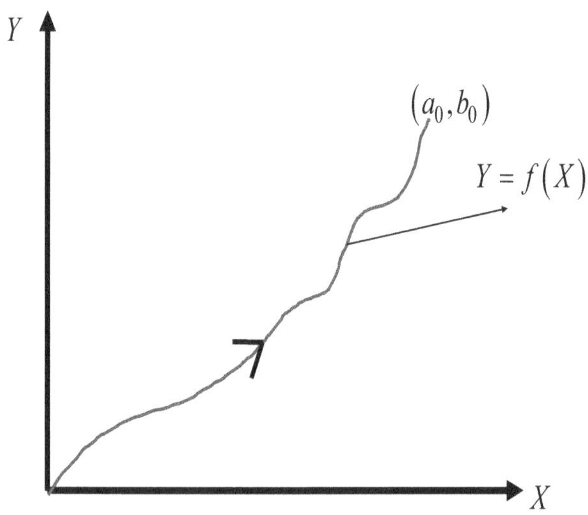

SOLUTION

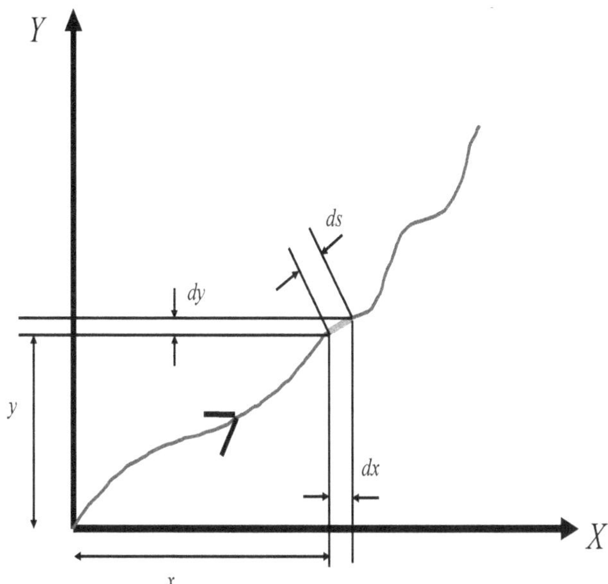

The differential magnetic force vector on the differential element *ds* shown is given by

$$d\vec{F} = I\left(d\vec{s} \times \vec{B}\right)$$

$$\Rightarrow d\vec{F} = I\left(dx\hat{i} + dy\hat{j}\right) \times \left(B_1\hat{i} + B_2\hat{j}\right) \tag{1}$$

Simplifying

$$d\vec{F} = I\left(B_2 \cdot dx - B_1 \cdot dy\right)\hat{k} \tag{2}$$

As per the question

$$\int_0^x I\left(B_2 - B_1 \cdot \frac{dy}{dx}\right)dx = a\frac{dy}{dx} \tag{3}$$

Differentiating both sides with respect to *x* (for the left-hand side, Leibnitz rule for differentiating a definite integral has been used)

$$I\left(B_2 - B_1 \cdot \frac{dy}{dx}\right) = a\left(\frac{d^2 y}{dx^2}\right)$$

$$\Rightarrow a\frac{d^2 y}{dx^2} + IB_1\frac{dy}{dx} - IB_2 = 0$$

(4)

The general solution to equation 4 is given by

$$y = C_1 e^{s_1 x} + C_2 e^{s_2 x}$$

(5)

Where

$$s_1 = \frac{-IB_1 + \sqrt{(IB_1)^2 + 4aIB_2}}{2a}$$

$$s_2 = \frac{-IB_1 - \sqrt{(IB_1)^2 + 4aIB_2}}{2a}$$

and C_1, C_2 are arbitrary integration constants. The boundary conditions needed to evaluate these constants are $y(0) = 0, y(a_0) = b_0$ .These yield

$$C_1 + C_2 = 0$$

(6)

$$C_1 e^{s_1 a_0} + C_2 e^{s_2 a_0} = b_0$$

(7)

Solving

$$C_1 = \frac{b_0}{e^{s_1 a_0} - e^{s_2 a_0}}$$

$$C_2 = -\frac{b_0}{e^{s_1 a_0} - e^{s_2 a_0}}$$

(8)

Feeding equation 8 in equation 5

$$y = \frac{b_0 \left(e^{s_1 x} - e^{s_2 x} \right)}{e^{s_1 a_0} - e^{s_2 a_0}} \tag{9}$$

Expanding back s_1, s_2

$$y = \frac{b_0 \left[e^{\left(\frac{-IB_1 + \sqrt{(IB_1)^2 + 4aIB_2}}{2a} \right) x} - e^{\left(\frac{-IB_1 - \sqrt{(IB_1)^2 + 4aIB_2}}{2a} \right) x} \right]}{e^{\left(\frac{-IB_1 + \sqrt{(IB_1)^2 + 4aIB_2}}{2a} \right) a_0} - e^{\left(\frac{-IB_1 - \sqrt{(IB_1)^2 + 4aIB_2}}{2a} \right) a_0}} \tag{10}$$

A particle of charge q and mass m rests at the origin. At $t=0$, it begins to move in the XY plane along the shown path. It moves in a manner such that the magnetic field produced by its motion at the point directly below it on the x axis at any instant is constant, equal to B. The magnitude of instantaneous velocity vector is proportional to the instantaneous y coordinate (proportionality constant being a). The permeability of vacuum is μ_0. Find the function f.

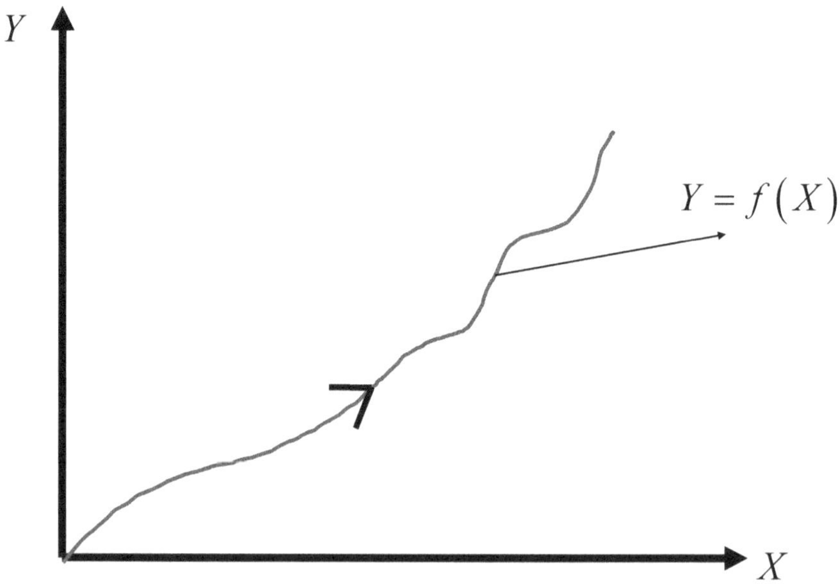

SOLUTION

The magnetic field vector produced by the moving charge at coordinates (x,y) is given by

$$\vec{B} = \frac{\mu_0 q \vec{v} \times \vec{r}}{4\pi |\vec{r}|^3} \tag{1}$$

$$\vec{B} = \frac{\mu_0 q \left(\frac{dx}{dt}\hat{i} + \frac{dy}{dt}\hat{j} \right) \times \left[\left(x\hat{i} \right) - \left(x\hat{i} + y\hat{j} \right) \right]}{4\pi \left| \left(x\hat{i} \right) - \left(x\hat{i} + y\hat{j} \right) \right|^3} \tag{2}$$

Simplifying

$$|\vec{B}| = \frac{\mu_0 q}{4\pi y^2} \frac{dx}{dt}$$

$$\Rightarrow B = \frac{\mu_0 q}{4\pi y^2} \frac{dx}{dt} \tag{3}$$

The other condition mentioned can be mathematically put as

$$\left(\frac{dx}{dt} \right)^2 + \left(\frac{dy}{dt} \right)^2 = a^2 y^2 \tag{4}$$

Equation 4 can be expressed as

$$\left(\frac{dx}{dt} \right) \sqrt{1 + \left(\frac{dy}{dx} \right)^2} = ay \tag{5}$$

Feeding $\dfrac{dx}{dt}$ from equation 5 in equation 3

$$B = \frac{\mu_0 q a}{4\pi y \sqrt{1 + \left(\frac{dy}{dx} \right)^2}} \tag{6}$$

Rearranging

$$\frac{dy}{dx} = \sqrt{\left(\frac{\mu_0 qa}{4\pi yB}\right)^2 - 1} \tag{7}$$

Let $\dfrac{\mu_0 qa}{4\pi B}$ be called c. Separating the variables and integrating both sides

$$\int_0^y \frac{y \cdot dy}{\sqrt{c^2 - y^2}} = \int_0^x dx \tag{8}$$

Let $c^2 - y^2 = U^2 \Rightarrow -y \cdot dy = U \cdot dU$

$$-\int_c^{\sqrt{c^2-y^2}} dU = \int_0^x dx \tag{9}$$

$$\Rightarrow \sqrt{c^2 - y^2} = c - x$$

Rearranging

$$y = \sqrt{2cx - x^2} \tag{10}$$

Expanding back c

$$y = \sqrt{\frac{\mu_0 qax}{2\pi B} - x^2} \tag{11}$$

Question 27

A particle of charge q and mass m rests at the origin. At $t=0$, it begins executing simple harmonic motion along the y axis, with amplitude A and angular frequency w. Find the instantaneous magnetic flux produced by this moving charge across the shown planar area lying in the XY plane. Permeability of free space is μ_0

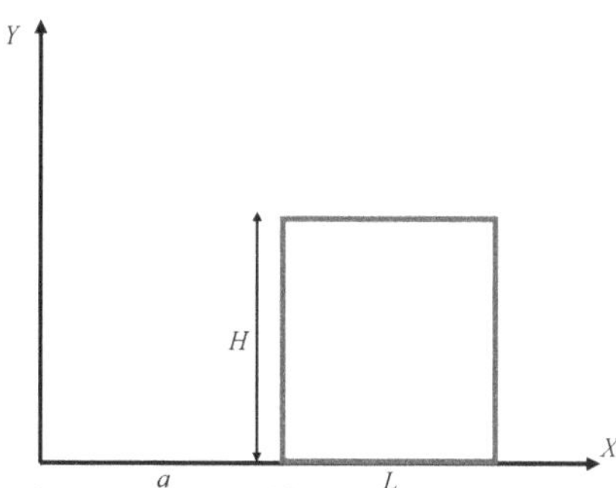

SOLUTION

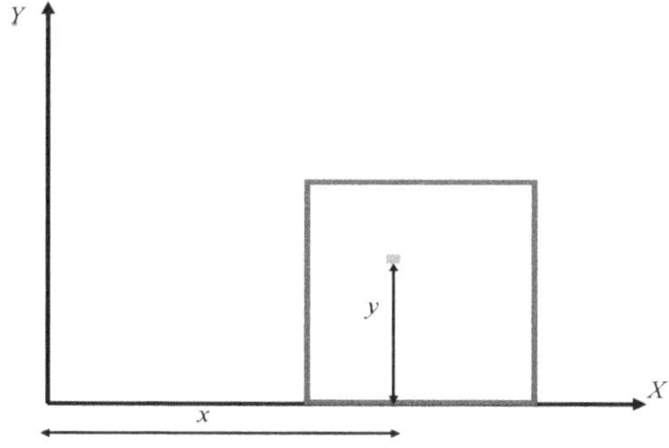

The magnetic field vector produced by the moving charge at time *t*, at the shown, general location inside the loop is

$$\vec{B} = \frac{\mu_0 q \vec{v} \times \vec{r}}{4\pi |\vec{r}|^3} \tag{1}$$

Where

$\vec{r}$ = position vector of location where field is needed-position vector of charge\

$$\Rightarrow \vec{r} = \left(x\hat{i} + y\hat{j} \right) - \left\{ A\sin(wt)\hat{j} \right\} \text{ and}$$

$$\vec{v} = \frac{d\left\{ A\sin(wt)\hat{j} \right\}}{dt}$$

$$\Rightarrow \vec{v} = Aw\cos(wt)\hat{j}$$

$$\vec{B} = \frac{\mu_0 q \left[Aw\cos(wt)\hat{j} \right] \times \left[x\hat{i} + y\hat{j} - A\sin(wt)\hat{j} \right]}{4\pi \left| x\hat{i} + y\hat{j} - A\sin(wt)\hat{j} \right|^3} \tag{2}$$

Simplifying

$$\vec{B} = -\frac{\mu_0 q \left[Awx\cos(wt)\hat{k} \right]}{4\pi \left[x^2 + \left\{ y - A\sin(wt) \right\}^2 \right]^{\frac{3}{2}}} \tag{3}$$

The magnetic flux across the shown differential area is

$$d\phi = \vec{B}.d\vec{A}$$

$$\Rightarrow d\phi = -\frac{\mu_0 q \left[Awx\cos(wt)\hat{k} \right]}{4\pi \left[x^2 + \left\{ y - A\sin(wt) \right\}^2 \right]^{\frac{3}{2}}} \cdot (dx \cdot dy)\hat{k} \tag{4}$$

The net magnetic flux across the region is given by

$$\phi = \int d\phi$$

$$\Rightarrow \phi = \int\limits_{a}^{a+L} \int\limits_{0}^{H} -\frac{\mu_0 q\left[Awx\cos(wt)\right]dx\cdot dy}{4\pi\left[x^2 + \{y\text{-}A\sin(wt)\}^2\right]^{\frac{3}{2}}} \tag{5}$$

Simplifying

$$\phi = -\frac{\mu_0 qAw\cos(wt)}{4\pi} \int\limits_{a}^{a+L} \int\limits_{0}^{H} \frac{x\cdot dx\cdot dy}{\left[x^2 + \{y\text{-}A\sin(wt)\}^2\right]^{\frac{3}{2}}} \tag{6}$$

Let $y\text{-}A\sin(wt) = x\tan\theta \Rightarrow dy = x\sec^2\theta\cdot d\theta$

$$\phi = -\frac{\mu_0 qAw\cos(wt)}{4\pi} \int\limits_{a}^{a+L} \frac{dx}{x} \int\limits_{\tan^{-1}\left[\frac{-A\sin(wt)}{x}\right]}^{\tan^{-1}\left[\frac{H-A\sin(wt)}{x}\right]} \cos\theta\cdot d\theta$$

$$\Rightarrow \phi = -\frac{\mu_0 qAw\cos(wt)}{4\pi} \int\limits_{a}^{a+L} \frac{dx}{x}\left[\sin\theta\right]_{\tan^{-1}\left[\frac{-A\sin(wt)}{x}\right]}^{\tan^{-1}\left[\frac{H-A\sin(wt)}{x}\right]} \tag{7}$$

$$\phi = -\frac{\mu_0 qAw\cos(wt)}{4\pi} \int\limits_{a}^{a+L}\left[\frac{H-A\sin(wt)}{\sqrt{x^2+\{H-A\sin(wt)\}^2}} + \frac{A\sin(wt)}{\sqrt{x^2+\{A\sin(wt)\}^2}}\right]\frac{dx}{x} \tag{8}$$

Let $x = \dfrac{1}{U} \Rightarrow dx = -\dfrac{dU}{U^2}$

$$\phi = \frac{\mu_0 qAw\cos(wt)\{H-A\sin(wt)\}}{4\pi} \int\limits_{\frac{1}{a}}^{\frac{1}{a+L}} \frac{dU}{\sqrt{1+U^2\{H-A\sin(wt)\}^2}}$$

$$+ \frac{\mu_0 qA^2 w\cos(wt)\sin(wt)}{4\pi} \int\limits_{\frac{1}{a}}^{\frac{1}{a+L}} \frac{dU}{\sqrt{1+U^2\{A\sin(wt)\}^2}} \tag{9}$$

$$\phi = \frac{\mu_0 qAw\cos(wt)\{H - A\sin(wt)\}}{4\pi} \int\limits_{\frac{1}{a}}^{\frac{1}{a+L}} \frac{dU}{\{H - A\sin(wt)\}\sqrt{U^2 + \dfrac{1}{\{H - A\sin(wt)\}^2}}}$$

$$+ \frac{\mu_0 qA^2 w\cos(wt)\sin(wt)}{4\pi} \int\limits_{\frac{1}{a}}^{\frac{1}{a+L}} \frac{dU}{A\sin(wt)\sqrt{U^2 + \dfrac{1}{\{A\sin(wt)\}^2}}} \tag{10}$$

$$\phi = \frac{\mu_0 qAw\cos(wt)}{4\pi} \ln\left| \frac{\dfrac{1}{a+L} + \sqrt{\dfrac{1}{(a+L)^2} + \dfrac{1}{\{H - A\sin(wt)\}^2}}}{\dfrac{1}{a} + \sqrt{\dfrac{1}{a^2} + \dfrac{1}{\{H - A\sin(wt)\}^2}}} \right|$$

$$+ \frac{\mu_0 qAw\cos(wt)}{4\pi} \ln\left| \frac{\dfrac{1}{a+L} + \sqrt{\dfrac{1}{(a+L)^2} + \dfrac{1}{\{A\sin(wt)\}^2}}}{\dfrac{1}{a} + \sqrt{\dfrac{1}{a^2} + \dfrac{1}{\{A\sin(wt)\}^2}}} \right| \tag{11}$$

Question 28

❖ ❖ ❖

A uniformly charged solid sphere of radius R , carrying a total charge Q rotates at an angular velocity w. Find magnetic field at a point outside it, at a distance H from its centre as shown. Permeability of free space is μ_0

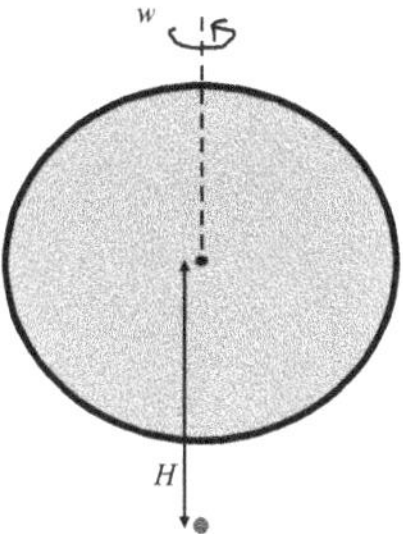

SOLUTION

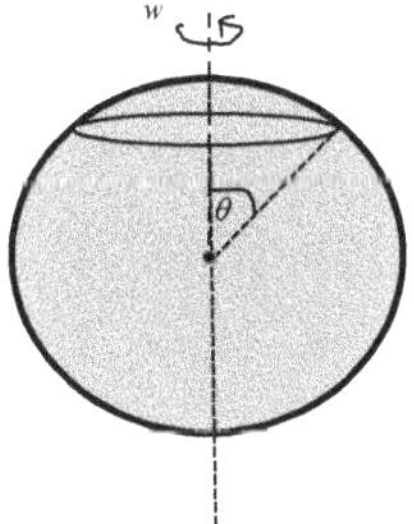

The magnetic field at a general location on the axis of a uniformly charged solid disc rotating about this axis is a standard result and is given by

$$B_d = \frac{\mu_0 \sigma w}{2}\left[\frac{r_d^2 + 2x^2}{\sqrt{r_d^2 + x^2}} - 2x\right] \tag{1}$$

Where σ is charge per unit area on the disc, r_d is radius of the disc, x is distance of the point on the disc's axis (where field is required) from the disc centre, w is angular velocity of the disc.

The sphere can be decomposed into differential discs as shown. The magnetic field of all such differential discs will be similarly directed. Hence, the overall field due to sphere shall be the integral of the field due to one such general disc. The areal charge density for the shown disc will be its charge per unit of its area, and its charge shall be total sphere charge per unit sphere volume multiplied by disc volume.

$$\sigma = \frac{Q}{\dfrac{4}{3}\pi R^3}\frac{\pi R^2 \sin^2\theta \cdot R\sin\theta \cdot d\theta}{\pi R^2 \sin^2\theta} \tag{2}$$

$$\Rightarrow \sigma = \frac{3Q\sin\theta \cdot d\theta}{4\pi R^2}$$

The axial distance of the given location from the shown disc's centre will be

$$x = R\cos\theta + H \tag{3}$$

Feeding equations 2 and 3 in equation 1

$$dB = \frac{\mu_0 w}{2}\frac{3Q\sin\theta \cdot d\theta}{4\pi R^2}\left[\frac{R^2\sin^2\theta + 2(R\cos\theta + H)^2}{\sqrt{R^2\sin^2\theta + (R\cos\theta + H)^2}} - 2(R\cos\theta + H)\right] \tag{4}$$

The net field is therefore

$$B = \int dB$$

$$\Rightarrow B = \frac{3Q\mu_0 w}{8\pi R^2} \int_0^\pi \left[\frac{R^2 \sin^2\theta + 2R^2\cos^2\theta + 2H^2 + 4RH\cos\theta}{\sqrt{R^2 + H^2 + 2RH\cos\theta}} - 2(R\cos\theta + H) \right] \sin\theta \cdot d\theta \qquad (5)$$

Let

$$R^2 + H^2 + 2RH\cos\theta = U^2$$

$$\Rightarrow \cos\theta = \frac{U^2 - R^2 - H^2}{2RH} \quad \text{and}$$

$$-\sin\theta \cdot d\theta = \frac{U \cdot dU}{RH}$$

$$B = -\frac{3Q\mu_0 w}{8\pi R^3 H} \int_{R+H}^{R-H} \left[\left(\frac{U^2 - R^2 - H^2}{2H} \right)^2 + 2U^2 - R^2 - \left(\frac{U^3 - UR^2 + UH^2}{H} \right) \right] dU \qquad (6)$$

Simplifying

$$B = -\frac{3Q\mu_0 w}{8\pi R^3 H} \int_{R+H}^{R-H} \left[\frac{U^4 + \left(R^2 + H^2\right)^2 - 2U^2\left(R^2 + H^2\right)}{4H^2} + 2U^2 - R^2 - \frac{U^3}{H} + \frac{U\left(R^2 - H^2\right)}{H} \right] dU \qquad (7)$$

$$B = -\frac{3Q\mu_0 w}{8\pi R^3 H} \int_{R+H}^{R-H} \left[\frac{U^4 - 4HU^3 + \left(6H^2 - 2R^2\right)U^2 + 4HU\left(R^2 - H^2\right) + \left(R^2 + H^2\right)^2 - 4H^2R^2}{4H^2} \right] dU \qquad (8)$$

$$B = -\frac{3Q\mu_0 w}{32\pi R^3 H^3} \left[\frac{U^5}{5} - HU^4 + \left(6H^2 - 2R^2\right)\frac{U^3}{3} + 2HU^2\left(R^2 - H^2\right) + \left(R^2 - H^2\right)^2 U \right]_{R+H}^{R-H} \qquad (9)$$

$$B = \frac{3Q\mu_0 w}{32\pi R^3 H^3} \left[\begin{array}{l} \left\{ \dfrac{(R+H)^5 - (R-H)^5}{5} \right\} - \\[2ex] H\left\{ (R+H)^4 - (R-H)^4 \right\} + \\[2ex] \left(6H^2 - 2R^2\right)\left\{ \dfrac{(R+H)^3 - (R-H)^3}{3} \right\} + \\[2ex] 2H\left(R^2 - H^2\right)\left\{ (R+H)^2 - (R-H)^2 \right\} + \\[2ex] 2H\left(R^2 - H^2\right)^2 \end{array} \right] \tag{10}$$

Question 29

❖ ❖ ❖

Aparticle of charge q and mass m is fixed above a hollow hemisphere of radius R, along its axis as shown. Find the electric flux produced by this charge across the hemisphere. The permittivity of free space is ε_0

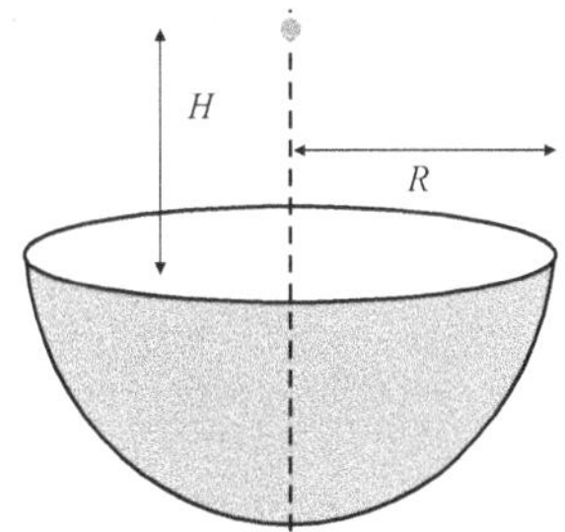

SOLUTION

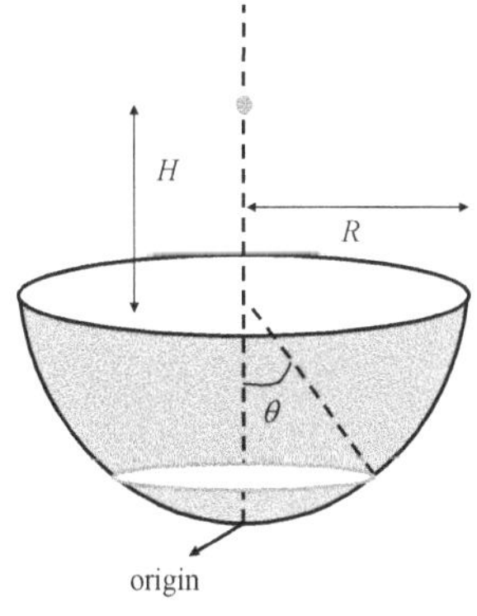

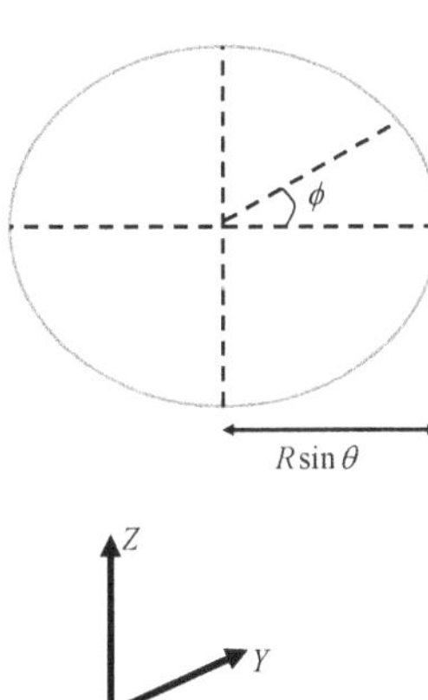

The distance between the point charge and the differential area shown is

$$D = \sqrt{\left(R\sin\theta\cos\phi - 0\right)^2 + \left(R\sin\theta\sin\phi - 0\right)^2 + \left[\left(R - R\cos\theta\right) - \left(R + H\right)\right]^2} \qquad (1)$$

Simplifying

$$D = \sqrt{R^2 + H^2 + 2RH\cos\theta} \qquad (2)$$

The magnitude of the electric field vector at this little area is

$$dE = \frac{q}{4\pi\varepsilon_0 D^2}$$

$$\Rightarrow dE = \frac{q}{4\pi\varepsilon_0 \left(R^2 + H^2 + 2RH\cos\theta\right)} \qquad (3)$$

The magnitude of the differential area shown is

$$dA = \left(R\sin\theta \cdot d\phi\right)\left(R \cdot d\theta\right) \qquad (4)$$

The differential area vector is along the local sphere radius vector. Therefore, the angle between the differential area vector and differential field vector will be the same as that between local sphere radius vector and differential field vector. This angle α is thus the angle between the following 2 lines:

The line joining $\left(R\sin\theta\cos\phi, R\sin\theta\sin\phi, R - R\cos\theta\right)$ and $\left(0,0,R+H\right)$

The line joining $\left(R\sin\theta\cos\phi, R\sin\theta\sin\phi, R - R\cos\theta\right)$ and $\left(0,0,R\right)$

$$\cos\alpha = \frac{\vec{b}_1 \cdot \vec{b}_2}{\left|\vec{b}_1\right|\left|\vec{b}_2\right|} \qquad (5)$$

where

$$\vec{b}_1 = \left[\left(R\sin\theta\cos\phi \right)\hat{i} + \left(R\sin\theta\sin\phi \right)\hat{j} - \left(H + R\cos\theta \right)\hat{k} \right]$$
$$\vec{b}_2 = \left[\left(R\sin\theta\cos\phi \right)\hat{i} + \left(R\sin\theta\sin\phi \right)\hat{j} - \left(R\cos\theta \right)\hat{k} \right]$$

(6)

$$\cos\alpha = \frac{R^2 + RH\cos\theta}{R\sqrt{R^2 + H^2 + 2RH\cos\theta}}$$
$$\Rightarrow \cos\alpha = \frac{R + H\cos\theta}{\sqrt{R^2 + H^2 + 2RH\cos\theta}}$$

(7)

The differential flux across the differential area dA is

$$d\phi = dE \cdot dA \cdot \cos\alpha$$

(8)

Feeding equations 3,4,7 in equation 8

$$d\phi = \frac{qR^2 \left(R + H\cos\theta \right)\sin\theta \cdot d\phi \cdot d\theta}{4\pi\varepsilon_0 \left(R^2 + H^2 + 2RH\cos\theta \right)^{\frac{3}{2}}}$$

(9)

The overall flux is

$$\phi = \int d\phi$$

$$\Rightarrow \phi = \int_0^{\frac{\pi}{2}} \int_0^{2\pi} \frac{qR^2 \left(R + H\cos\theta \right)\sin\theta \cdot d\theta \cdot d\phi}{4\pi\varepsilon_0 \left(R^2 + H^2 + 2RH\cos\theta \right)^{\frac{3}{2}}}$$

(10)

$$\Rightarrow \phi = \int_0^{\frac{\pi}{2}} \frac{qR^2 \left(R + H\cos\theta \right)\sin\theta \cdot d\theta}{2\varepsilon_0 \left(R^2 + H^2 + 2RH\cos\theta \right)^{\frac{3}{2}}}$$

Let

$$R^2 + H^2 + 2RH\cos\theta = U^2$$

$$\Rightarrow \cos\theta = \frac{U^2 - R^2 - H^2}{2RH} \quad \text{and}$$

$$-\sin\theta \cdot d\theta = \frac{U \cdot dU}{RH}$$

$$\phi = -\frac{q}{4H\varepsilon_0} \int_{R+H}^{\sqrt{R^2+H^2}} \left(1 + \frac{R^2 - H^2}{U^2}\right) dU \tag{11}$$

$$\phi = -\frac{q}{4H\varepsilon_0} \left[U - \frac{\left(R^2 - H^2\right)}{U}\right]_{R+H}^{\sqrt{R^2+H^2}} \tag{12}$$

$$\Rightarrow \phi = \frac{q}{2\varepsilon_0} \left[1 - \frac{H}{\sqrt{R^2 + H^2}}\right]$$

Question 30

A particle of charge Q_0 is fixed at the centre of a hollow hemisphere of radius R. Another particle of charge q and mass m rests at the bottom of the hemisphere. The friction coefficient between this charge and hemisphere is μ. At $t=0$, this particle is imparted a tangential velocity v_0 and simultaneously, the fixed particle begins to lose charge at a constant rate r. Find the time it takes for the moving particle to come to its original position again. Neglect gravitational, air resistance and buoyant forces. The permittivity of free space is ε_0

SOLUTION

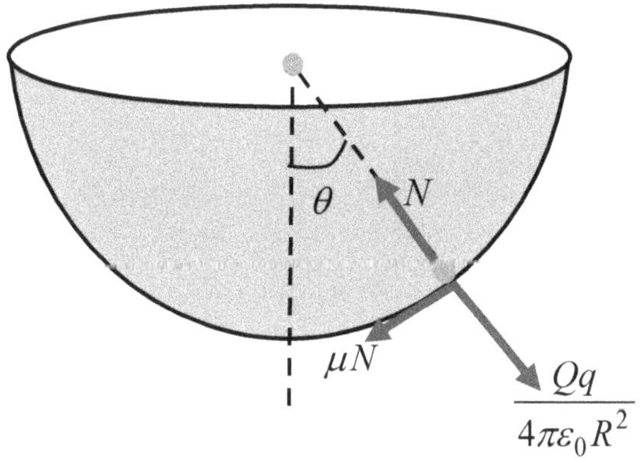

The instantaneous charge carried by the fixed particle is

$$Q = Q_0 - rt \tag{1}$$

Applying Newton's second law of motion for the moving particle at this general time t, in the radial and tangential directions

$$N = \frac{Qq}{4\pi\varepsilon_0 R^2} \tag{2}$$

$$-\mu N = m\frac{dv}{dt} \tag{3}$$

Eliminating N from equations 2 and 3, and feeding equation 1 in the resulting equation

$$-\frac{\mu q (Q_0 - rt)}{4\pi\varepsilon_0 R^2} = m\frac{dv}{dt} \tag{4}$$

Separating the variables and integrating both sides

$$\int_{v_0}^{v} dv = -\int_{0}^{t} \frac{\mu q (Q_0 - rt)\, dt}{4\pi\varepsilon_0 m R^2}$$

$$\Rightarrow v = v_0 - \frac{\mu q \left(Q_0 t - \dfrac{rt^2}{2} \right)}{4\pi m \varepsilon_0 R^2} \tag{5}$$

Using differential definition of velocity

$$R\frac{d\theta}{dt} = v_0 - \frac{\mu q \left(Q_0 t - \dfrac{rt^2}{2} \right)}{4\pi m \varepsilon_0 R^2}$$

$$\Rightarrow \frac{d\theta}{dt} = \frac{v_0}{R} - \frac{\mu q \left(Q_0 t - \dfrac{rt^2}{2} \right)}{4\pi m \varepsilon_0 R^3} \tag{6}$$

Separating the variables and integrating both sides

$$\int_0^\theta d\theta = \int_0^t \left[\frac{v_0}{R} - \frac{\mu q\left(Q_0 t - \dfrac{rt^2}{2} \right)}{4\pi m\varepsilon_0 R^3} \right] dt$$

$$(7)$$

$$\Rightarrow \theta = \frac{v_0 t}{R} - \frac{\mu q\left(\dfrac{Q_0 t^2}{2} - \dfrac{rt^3}{6} \right)}{4\pi m\varepsilon_0 R^3}$$

The required times at which θ is 0, other than $t=0$ are given by

$$t_0 = \frac{\dfrac{\mu q Q_0}{8\pi m\varepsilon_0 R^3} \pm \sqrt{\left(\dfrac{\mu q Q_0}{8\pi m\varepsilon_0 R^3} \right)^2 - \dfrac{v_0 \mu q r}{6\pi m\varepsilon_0 R^4}}}{\dfrac{\mu q r}{12\pi m\varepsilon_0 R^3}}$$

$$(8)$$